THE
LIFE
ERA

Books by Eric Chaisson

COSMIC DAWN: *The Origins of
Matter and Life*

THE INVISIBLE UNIVERSE: *Probing
the Frontiers of Astrophysics*
(coauthored with George Field)

THE LIFE ERA: *Cosmic Selection
and Conscious Evolution*

Eric Chaisson

Illustrated by

Lola Judith Chaisson

THE
LIFE
ERA

Cosmic

Selection

and

Conscious

Evolution

The Atlantic
Monthly Press
New York

The Life Era is a volume in the Evolutionary Synthesis Series of the General Evolution Research Group.

FIRST EDITION

Library of Congress Cataloging-in-Publication Data

Chaisson, Eric.
  The life era.

  The present volume is the second entry of a planned trilogy of semi-technical works on cosmic evolution.
    1. Cosmology. 2. Life—
Origin. I. Title.
QB981.C416 1987   523.1   86-32296
ISBN 0-87113-062-9

Published simultaneously in Canada
Printed in the United States of America

FIRST PRINTING

Design by Laura Hough

To Megan Lyra and Paul Cygnus,
among the vanguard of the Life Era

CONTENTS

THE
LIFE
ERA

PROLOGUE

Four
Objectives

> *The telescope at one end of his beat,*
> *And at the other end the microscope,*
> *Two instruments of nearly equal hope. . . .*
> *—Robert Frost*

WE ARE ENTERING an age of synthesis.

Tucked away in laboratories, working by night at observatories, and talking a mathematical language that regrettably few people can fathom, natural scientists are steadily unveiling some of the most basic mysteries of our Universe. Using astronomical telescopes and biological microscopes among a virtual arsenal of other tools of high technology, we have discovered a thread of understanding linking the origin and existence of all things. Now emerging is a unified scenario of the cosmos, including ourselves as sentient beings, based upon the time-honored concept of change. (Change: To make different the form, nature, and content of something.) From galaxies to snowflakes, from stars and planets to life itself, we are beginning to identify an underlying pattern penetrating the fabric of all the natural sciences—a sweepingly encompassing view of the formation,

1

structure, and maintenance of all objects in our richly endowed Universe.

I feel fortunate to be an astrophysicist at this stage in human history. I imagine that when our great-grandchildren gain perspective on the last portion of the twentieth century, they will likely conclude that we now share a golden age of astrophysics. I say this because the number and diversity of discoveries currently being made are—shall I risk it?—astronomical. In particular, we are currently exploring all the remaining parts of the electromagnetic spectrum, thereby granting us some early glimpses of invisible radiation, including radio, infrared, and ultraviolet waves, as well as X and gamma rays. In hardly more than a single generation—not the generation of our parents, and not that of our children, but *our* generation—astronomers are now revealing the invisible cosmos much as Galileo first sampled magnified light from visible astronomical objects. The result is unsurpassed intellectual excitement concerning the nature of the Universe as well as of our role in it.

At the same time I suggest that future historians will probably judge that we also now share a golden age of biochemistry. The rapid pace and penetrating insight of novel breakthroughs in the biological sciences in many ways equal the impressiveness of those of the physical sciences. The unraveling of life's code and the advent of genetic engineering, to cite but a couple of advances, herald a renewed vigor within the biochemical community.

Actually I am doubly fortunate, for in my research and teaching, I have contributed to each of the interdisciplines of astrophysics and biochemistry, and furthermore, I have recently been attempting to synthesize these two subjects into an even grander transdiscipline that I call cosmic evolution. Simply defined, *cosmic evolution is the study of the many varied changes in the assembly and composition of energy, matter, and life in the Universe.* In even briefer terms, cosmic evolution is a cosmology wherein life plays an integral role.

Now I realize that a considerable fraction of the world's populace, most notably in the United States, can become emotional, even irate, and occasionally convulsive at mention of the word "evolution." But let me assure them at the outset that evolution implies neither dogmatism nor atheism. Evolution is hardly more than a fancy word for change—especially developmental change. Indeed, it seems that change is the hallmark for the origin, development, and

2

. . . we are beginning to identify an underlying pattern penetrating the fabric of all the natural sciences . . .

3

maintenance of all things in the Universe, animate or inanimate. Change has, over the course of all time and throughout all space, brought forth, successfully and successively, galaxies, stars, planets, and life. Thus, we give this process of universal change a more elegant name—cosmic evolution, which for me includes all aspects of evolution: particulate, galactic, stellar, elemental, planetary, chemical, biological, and cultural. As such, the familiar subject of biological evolution becomes just one segment of a much broader evolutionary scheme stretching well beyond mere life on Earth. In short, what Darwin once did for plants and animals, cosmic evolution does for all things. And if Darwinism created a veritable revolution in understanding by helping to free us from the anthropocentric belief that humans basically differ from other life forms on our planet, then cosmic evolution is destined to extend that intellectual revolution by in turn releasing us from regarding matter on Earth and in our bodies any differently from that in the stars and galaxies beyond.

In effect, though I acknowledge its implied arrogance and pretentiousness (yet it's the most succinct description I can presently offer), with cosmic evolution as the core, we are trying to create a new philosophy—a scientific philosophy. And I hasten to emphasize the adjective "scientific," for unlike classical philosophy, observation and experimentation are key features of this new effort; to be sure, I wholeheartedly subscribe to the notion that neither thought alone nor belief alone will ever make the unknown known. Cosmic evolution is designed to address the fundamental and age-old questions that philosophers and theologians have traditionally asked, but to do so using the scientific method and especially the instruments of state-of-the-art technology.

Indeed, the same technology that threatens to doom us now stands ready to probe meaningfully some of the most basic issues: Who are we? Where did we come from? How did everything around us, on Earth and in the Universe, originate? What is the source of order, form, and structure characterizing all things material? What are their destinies? Of ultimate import, armed with a renewed and quantified perception of change, some scientists now seem poised to tackle the fundamentally fundamental query—to wit, Why is there something rather than nothing?

*   *   *

In my first book on this subject, *Cosmic Dawn: The Origins of Matter and Life*, I sketched without using mathematics the salient features of cosmic evolution. Employing time as the (not surprisingly) underlying theme for mapping the consequences of change, I therein outlined the myriad and ongoing events that likely gave rise to galaxies, stars, planets, and selves. I stressed how the simple concept of change, often manifesting itself in subtle yet complex ways, guides the evolution of the Universe and all its contents—in particular how matter naturally and of its own accord materialized from primordial energy of the early Universe, and how life in turn likewise emerges, given sufficient time and resources, from that matter. Toward the end of *Cosmic Dawn*, I broached the notion of a universal Life Era at whose threshold we now reside on planet Earth, thus positing (without scientific testimony) that it is the existence of curious, technologically intelligent beings (here on Earth and perhaps elsewhere) that grants meaning to the cosmos in the guise of an evolved universal consciousness.

In the intervening several years since I authored *Cosmic Dawn*, numerous thoughtful readers have inquired further concerning four major aspects of cosmic evolution and of its implications:

—How old is the notion of cosmic evolution, and how has the idea of change itself changed over the course of time?

—How can order emerge from chaos, and especially complex life from simple chemicals, when the second law of thermodynamics dictates that the Universe steadily becomes increasingly randomized and disordered?

—How did the Universe originate; in particular, what is the origin of the primal energy at creation?

—How does my view of the Life Era compare to previous efforts by philosophers and theologians to build grand schemes of life's destiny, and what are its implications for the future of our human species?

In the present volume, the second entry of a planned trilogy of semitechnical works on cosmic evolution, I specifically address each of these areas of concern. After outlining in Chapter 1 the scenario of cosmic evolution (which is but an anecdotal précis of

Cosmic Dawn), I trace in Chapter 2 the history of the idea of change. This basic concept is deeply rooted in our intellectual heritage, extending back at least twenty-five centuries to the time of the pre-Socratics. But evolution's rich history in no way cheapens current efforts to decipher the ways and means of change, for I stress again (without appeal to anthropocentrism) that only recently have we invented the equipment to help transform an ancient classical philosophy into a modern scientific philosophy.

In Chapter 3, I summarize efforts now under way to explain the substantial physical and biological order of our world, despite the popular (but incorrect) impression that the principles of thermodynamics unwaveringly work against the establishment of order. Of distinct notability, the existence of complex structures and especially of life itself does not violate the second law of thermodynamics. In this regard, we seem in the midst of a paradigmatic shift whereby the concepts of irreversible time, statistical probability, and chance are replacing the older, deterministic, even mechanistic attitude of Newtonian physics.

I discuss in Chapter 4 the early Universe, including the recent and provocative theory that the cosmos might well have self-originated from literally nothing—a mere quantum fluctuation in a vacuum having zero energy. I also describe there the two preeminent changes of all time—namely, the emergence of matter from energy and, in turn, that of life from matter. For technically inclined readers, a mathematical appendix is included in quantitative support of this greatest pair of all transformations.

The last chapter explores the implications for the Life Era, a future epoch of opportunity that we ourselves have helped to create but one that we can even more easily deny our children and all those humans who would follow. I make a point of contrasting this era's highlights with several similarly grandiose schemes derived largely via philosophical and theological inquiry. And while it may embody demonstrably nonscientific overtones, my view of the Life Era derives strictly from the principles of modern science. It is on this, in my opinion more natural (i.e., not supernatural) basis that I reason the Life Era comprises a third grand period in the history of the Universe.

As a crucial aspect of this last chapter, I urge the adoption of a global, perhaps even cosmic set of ethics if we on Earth are suc-

6

. . . a future epoch . . . we can . . . easily deny our children . . .

cessfully to enter the Life Era. I suggest (as I will more fully argue in the last volume of the intended trilogy) that we must now develop an integrated worldly culture, including a unified politico-economic ideology—which is not just a hackneyed proposal for a world government—if we as a species are to have a future. What's more, only by thinking big and embracing change can we ensure the onset of a planetwide program of ethics. Indeed, if we act wisely, quite beyond merely intelligently, then an epoch of something resembling "ethical evolution" should naturally emerge as the next great evolutionary leap forward in the overall scheme of cosmic evolution itself. Yet it is an evolutionary advance dependent upon technologically talented life forms—namely, us, the new agents of change—to plan and implement, lest we succumb to the seeds of our own destruction.

In general though straightforward terms, as I see it, those technological civilizations (on whatever planet) that do succeed in appreciating the import of cosmic evolution sufficiently to embrace the absolute need for a global ethics will survive, and those that don't won't. Life in the Universe is probably governed by a principle of cosmic selection—a sort of natural selection operating throughout the grandest realms of space and time. Only those life forms welcoming, even seeking, planetary citizenship as part of a unitary globalism will likely realize the Life Era.

Finally, as a gesture toward the unknown (though of course here I am less sure, even metaphysical), the Epilogue addresses the potential evolution of universal consciousness via the self-nourishment of its sentient beings, thus enabling us to decipher cosmological meaning and purpose from the amalgamated perspective of science *and* philosophy.

As in *Cosmic Dawn,* I have forgone citing living researchers (unless they are quoted directly) in favor of literary coherence and readability. To cite the legions of workers now contributing to the numerous subdisciplines comprising the grand scenario of cosmic evolution not only would risk slighting those I might omit but more crucially would likely detract from the clarity of the conceptual arguments stressed throughout. At the end of the book I have provided a representative bibliography of some of the many fine works I found useful while addressing the topics of this book. The reader should find these references appropriate for further study.

I gratefully acknowledge the support of my wife, Lola, who has instrumentally shaped my semitechnical scientific exposition and who has herein drawn numerous thought-provoking illustrations often bridging the two cultures of science and art with the accuracy of the former and the pizzazz of the latter. I am also indebted to Upton Brady, Ervin Laszlo, Sarah McFall, David Schuman, Mark Stier, Christopher Stone, Wendy Strahm, and Sir Crispin Tickell for commenting on the manuscript, and to The Atlantic Monthly Press for creating a publishing relationship that would be the envy of most authors. My opinions expressed regarding the appropriate formula for the survival of our species, however, have not necessarily been endorsed by any of these readers or in any way whatsoever by the institutions with which I am currently affiliated—a good thing, for if all intelligent people agreed upon these grand issues, we'd likely expire of boredom, if not inadaptability. Lastly, I thank the students of the Harvard-Radcliffe and Haverford-Bryn Mawr bicollege communities for helping me develop a better appreciation for the historical and philosophical aspects of my research.

Eric J. Chaisson
Spring, 1986
Villanova, Pennsylvania

CHAPTER 1

Our Cosmic Heritage

POLYMATHS ARE NOW USING THE CONCEPT OF CHANGE TO BUILD A GRAND SYNTHESIS OF ALL ENERGY, MATTER, AND LIFE IN THE UNIVERSE

HUMANS NOW SHARE A GOLDEN AGE of exploration and understanding. We have gathered more facts during the past few decades than throughout all the nearly hundred centuries of recorded history. In the process, aided by technology, we are transferring some of the most basic and profound issues from the realms of philosophy and religion to the domain of science. The result is an unprecedentedly rich view of ourselves and our Universe.

Here's a smattering of knowledge acquired during roughly the past half century:

—The Universe has not existed forever but had a definite beginning about fifteen billion years ago.

—Most matter in the Universe is inherently dark, invisible to our human eyes even when aided by large optical telescopes.

11

—The Sun is not at the center of the Universe.

—All stars advance through cycles of birth, maturity, and death, much like life forms on planet Earth.

—Innumerable stars have perished to create the matter now composing our world.

—We ourselves are made of atoms fused in the hearts of stars.

—Life probably arises naturally from nonliving matter, given the proper ingredients, genial environment, and enough time.

—Large molecules, called genes, are responsible for the passage of heredity from one generation to another.

—Erect, tool-using humans have roamed Earth for the last few million years.

Knowledge has exploded to the point where some of us feel the time is right for a grand synthesis of science—a merger of the biological and physical disciplines among perhaps others, a unification of ourselves with the cosmos. In this way we can begin to appreciate how *all* things—from atoms to roses, from galaxies to people—are interrelated. Accordingly, we are now in the process of sketching an encyclopedic scenario termed cosmic evolution—a mixture of proved facts and testable ideas describing a long series of changes in the assembly and composition of energy, matter, and life in the Universe. These are the changes that have produced in turn our Galaxy, our Sun, our Earth, and ourselves.

The idea that "change dominates all" is not really a new one. Twenty-five centuries ago the Greek philosopher Heraclitus reportedly declared something to the effect that "There is nothing permanent except change." Hardly could he have then known that modern science would embrace his idea and back it with experimental tests. Indeed, the concept of change—especially developmental change or evolution—has become a central theme of twentieth-century research, the hallmark for the creation of all things.

I daresay that we can now trace a thread of understanding linking the evolution of primal energy into elementary particles, the

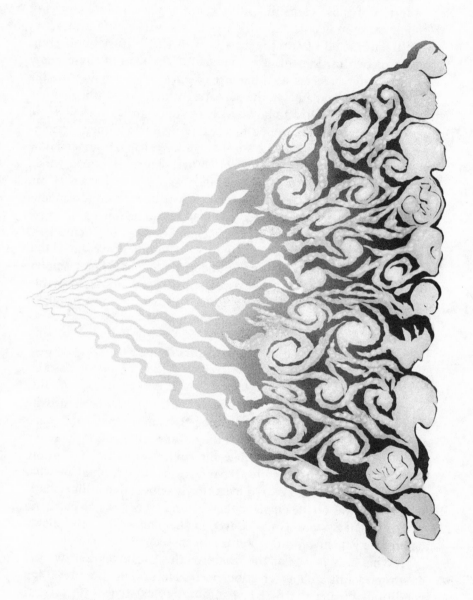

. . . a long series of changes in the assembly and composition of energy, matter, and life in the Universe.

evolution of those particles into atoms, in turn of those atoms into galaxies and stars, the evolution of stars into heavy elements, the evolution of those elements into the molecular building blocks of life, of those molecules into life itself, of advanced life forms into intelligence, and of intelligent life into the cultured and technological civilization that we now share. The sum of these many varied changes creates a comprehensive scenario coupling the big and small, the near and far, the past and future into a unified whole.

Why do we strive to build such a grand synthesis? Well, if we can understand the essence of cosmic evolution, then perhaps we can better appreciate who we are, where we came from, and how we fit into the cosmic scheme of things. We shall gain a wider, integrated knowledge of stars and galaxies, plants and animals, air, land, and sea. In particular, we shall learn how living organisms evolved the expertise to invade the land, generate language, create culture, devise science, explore space, and even study themselves.

In short, we are now striving to frame a heritage—a cosmic heritage—a sweeping structure of understanding based on events of the past, an intellectual road map identified and embraced by humans of the present, indeed a virtual blueprint for survival if adopted by our descendants of the future.

> Order is heav'n's first law.
> —Alexander Pope

To address the basic questions, we must probe far back into the past; beyond our birth dates some tens of years ago, beyond Renaissance times centuries ago, beyond the onset of civilization some ten thousand years ago. To appreciate the subject of cosmic evolution, we must broaden our horizons, expand our minds, and imagine what it was like long, long ago. Go back, for instance, five billion years, when there was no life on planet Earth; when there was no Earth, no Sun, no Solar System. Go back even farther. Before the stars shone, before the galaxies formed. Return to the beginning of time.

What was it like at the origin of the Universe? Can we say anything definite about creation itself? How about the prevailing conditions during the first few minutes of existence? To fathom events that occurred eons ago, astrophysicists are forced to rely on models. After all, times long past are times long gone. This doesn't mean that we just sit back, gaze out the window, and guess. As scientists we base our models on a wealth of data dictating the scale and dynamics of the Universe.

The emphasis on real data and testable ideas is an important one. As the British astronomer Sir Arthur Eddington declared, "For the truth of the conclusions of physical science, observation is the supreme court of appeals." Less ceremoniously his colleague Lord Kelvin once quipped, "Unless you have measured it, you don't know what you are talking about."

So, how do we measure the early Universe? We do so by studying the galaxies that light up the far away and the long ago. And by observing a weak radio hiss filling the whole Universe—the cooled relic of the aftermath of creation itself, the fossilized grandeur of a bygone era. This radio radiation seems to be omnipresent, for it surrounds all of us and is constantly absorbed by our bodies at a minuscule rate less than a millionth of the power of a hundred-watt light bulb.

One doesn't need to be a professional astronomer to appreciate some of the evidence for the early Universe. A good pair of binoculars permits anyone to see some nearby galaxies. And even some of the static on our home AM radios arose from ancient epochs of the Universe; obnoxiously crackling and hissing, some of that noise dates back to nearly the beginning of time.

I often liken studies of the early Universe to the efforts of bomb squads. Examining the debris from an explosion, police technicians try to reconstruct the nature of the bomb. In the case of modern astronomy the debris is the distant galaxies and the radio static, while the explosion was the origin of the Universe itself. Using, for example, the radically constructed Multiple-Mirror Telescope atop Arizona's Mount Hopkins to explore the galaxies hurtling away in our expanding Universe, or operating Bell Laboratories "Sugarscoop" horn antenna in northern New Jersey to measure the radio relic of ancient ages, we can glean some insight into the ultimate cosmic bomb—the Big Bang that started it all some fifteen billion years ago.

By most accounts, the Universe began with the explosion of something hot and dense—hotter than the tens of millions of degrees Celsius in the cores of most stars, denser than the trillions of grams per cubic centimeter in the nucleus of any atom. Precisely

15

what that "something" was, we cannot currently say with much certainty. Perhaps nothing more than a bolt of energy. Or perhaps nothing at all, as discussed in Chapter 4. And why that something exploded, we really don't know. The origin itself resembles the "Here there be dragons" school of ancient cartography. Still, people persist in asking, "What happened before the bang?" Frustrated, I often resort to St. Augustine, who long ago reportedly mused, "Before the Universe began, hell was being created for people who worry about such issues."

Imagine the Universe in the first few minutes of its being. The biggest of all possible explosions had just occurred. Energy flooded every available niche. And a whole mélange of exotic subatomic particles whizzed this way and that, surging amidst great heat and light. These particles did not come from thin air; in point of fact, there was no air at the time, thin or otherwise. The particles simply "materialized" from the energy of the primeval bang. Neither magic nor mysticism prevailed, just the well-known and oft-studied fact that matter can be naturally created from clashes among packets of energy. This interchangeability of matter, m, and energy, E, is proved daily in the bowels of nuclear reactors around the world. Together, they obey Albert Einstein's famous formula, $E = mc^2$, where c symbolizes the speed of light.

In those first few minutes of the Universe, energy reigned supreme, vaporizing all but the smallest chunks of matter. Protons, neutrons, and electrons, as well as a veritable zoo of other submicroscopic particles were unable to assemble into anything more complex. No stars or planets existed at the time. Not even any atoms were tolerated. The Universe was just too hot and chaotic— the clear and frenzied aftermath of a cosmic bomb.

With time's passage, the Universe changed rapidly. Of foremost importance, it cooled and thinned. Sometime between the first few minutes and the first million years—the nature of the physical process was gradual—the elementary particles of matter began clustering. Electrical forces bound the particles into atoms; the weakened energy could no longer break them apart. In effect, matter had gained some leverage over the previously dominating energy. *I regard this change from energy-dominance to matter-dominance as the first of two preeminent events in the history of the Universe.*

In this way, immense quantities of the simplest chemical element, hydrogen, were created in the early Universe; single protons captured single electrons. Some helium was also produced at the time, when hydrogen nuclei (protons) collided, stuck, and fused into heavier particles, later to be joined by a pair of electrons. But elements heavier than helium could not have been appreciably cooked in the early Universe. The fibers composing this page you are reading, the oxygen and nitrogen in the air we breathe, the copper and silver of the coins in our pockets were not fashioned in the aftermath of the initial explosion. The Universe simply cooled and thinned too rapidly to fabricate the heavy chemical elements.

With the relentless march of time, the Universe grew thinner, colder, and darker. It evolved much more slowly in later epochs, but it evolved nonetheless.

> *I could be bounded in a nutshell, and*
> *count myself a king of infinite space. . . .*
> —*William Shakespeare, Hamlet II, ii*

Deep space harbors myriad objects looking strangely unlike stars. Photographic time exposures, taken with even small, six-inch telescopes, often reveal fuzzy, lens-shaped images resembling disks more than the bright, round points of light we call stars. The eighteenth-century German philosopher Immanuel Kant regarded these flattened objects as "island universes" far beyond the confines of our own star system. I can't imagine there being more than one Universe, since the word's very definition is "the totality of all things." (Those people who even today speak of multiple universes pose a semantics problem that, to me, makes no sense scientifically.) But Kant was correct in arguing that these nonstellar patches of light reside in realms so far as to be imperceptible to the naked human eye.

Not until about a half century ago did Edwin Hubble use the 100-inch telescope on California's Mount Wilson to resolve these blurry and distant beacons into galaxies—huge structures of matter among the first to form after the Big Bang. Spanning on average a hundred thousand light-years across, galaxies might be better described as gargantuan; a single light-year equals the distance traveled by light in a full year—namely, about ten trillion kilometers. Unquestionably, galaxies are typical of the size and scale from which today's popular adjective "astronomical" derives.

17

Furthermore, with hundreds of billions of stars bound loosely by gravity, each galaxy houses more stars than people who have ever lived on Earth. Gargantuan, to be sure, for if you could exact a tax of a penny per star, a stellar census of a single galaxy would make you a billionaire. What's more, outer space is strewn with enough known galaxies to make every inhabitant on Earth that rich. Astronomical indeed.

Silently and majestically, galaxies twirl in the faraway tracts of the Universe—vast pinwheels of energy, matter, and perhaps life—imparting a feeling simultaneously for the immensity of the Universe and for the mediocrity of our position in it.

Floating proudly and mutely, our daytime Sun as well as all the nighttime stars share membership in one such galaxy—a huge assemblage of stars termed the Milky Way. It's hard to know exactly what our own Galaxy looks like. We can't get outside it and look back. We live inside the Milky Way, confined near the edge of this truly expansive system of stars. "We reside in the galactic suburbs," the American astronomer Harlow Shapley used to say.

Part of my research involves the exploration of wholly uncharted pieces of galactic real estate, especially the central regions of our Milky Way. Frustrated one evening at the Harvard Observatory, I wandered to Cambridge Common, where I perched myself on the back of a bench near the periphery of the park. Straining to fathom the locations of crosswalks, benches, and trees, I gained some insight into the problem. Barring myself from walking, bicycling, or otherwise sauntering about, I soon discovered that mapping the park's layout would be no easy task. Strange and intriguing objects—especially the grand monument near the park's center—seemed shrouded in mystery, for they resembled none of the familiar shrubs and benches near the edge of the park. And so it is with our Milky Way Galaxy. Relegated perhaps forever to the galactic boondocks, we currently strain to unravel the spread of stars, gas, and dust in that part of the Universe we call our galactic home. Not surprisingly, our view from the suburbs is limited.

Our work is further hampered because most of our Galaxy is invisible. Not even the world's largest optical telescopes atop America's Mount Palomar or Russia's Caucasus, which multiply our human vision a million times, can see much of it. Pound for pound, the Milky Way is a relatively dirty place, and light radiation just can't penetrate the vast accumulations of interstellar dust any more than it can illuminate roadside objects in a dense terrestrial fog. (If we could compress a typical parcel of interstellar space to equal the density of air on Earth, this parcel would contain enough dust to make a fog so thick we'd be lucky to see our hands held at arm's length in front of us.) Even so, for two decades now, radio and infrared researchers have probed the Galaxy for long-wavelength invisible radiation virtually immune to scattering by the galactic debris. And what we have found resembles a huge, flattened pinwheel of clouds and stars, some hundred thousand light-years across, bathed in an even larger, more spherical halo of thin, dark matter. The Sun and its system of planets reside about thirty thousand light-years from the center.

Normal galaxies, much like our spiral Milky Way, are spread throughout space at least as far away as several billion light-years. Beyond this distance roam countless more galaxylike objects, though clearly abnormal ones. Peculiarly, these distant beacons emit their radiation from paired lobes of plasma (hot, charged gas sometimes called the "fourth state of matter") that routinely span millions of light-years. Called active galaxies, objects beyond several billion light-years are more powerful, to some extent more violent than normal galaxies. Radiating hundreds, sometimes thousands of times more energy than our entire Milky Way, each of these distant objects seems to behave quite unlike collections of ordinary stars. To be truthful, we're unsure if active galaxies even have any stars.

Enigmatic though they may be, the active galaxies are not the most energetic objects in the Universe. For two decades now, we've known of astoundingly luminous and distant objects—astronomical bodies so puzzling that they threaten to topple the currently known laws of physics. These are the innocuous-looking, though incredibly powerful, quasi-stellar sources called quasars for short. Not content just to rival the bizarre properties of active galaxies, quasars actually extend those troubles. Here's why.

Quasar radiation often fluctuates from week to week, sometimes from day to day. Since cause-and-effect arguments demand that no light source flicker more quickly than radiation can cross it (lest the flickering we observe be blurred), the huge energies of the quasars must arise from regions tiny by cosmic standards; in other words, the whole of the quasar must change at about the same time

19

. . . a huge, flattened pinwheel of clouds and stars, some hundred thousand light-years across, bathed in an even larger, more spherical halo of thin, dark matter. (Arrow denotes the Sun's position.)

20

to preserve the coherency of its emitted light. How small must quasars typically be? Hardly much larger than our Solar System. Thus, our observations imply that quasars emit vast quantities of energy (often equivalent to about a thousand Milky Ways) while simultaneously being rather small in size. Therein lies perhaps the foremost dilemma in all of astrophysics: Quasars' large energies yet small dimensions seem incompatible, especially when we realize that their radiation is often launched at all wavelengths, from radio waves to X rays.

That the quasar paradox is strongly ingrained in the minds of many scientists can be perhaps appreciated by our reading a few lines from the pen of Caltech's Jesse Greenstein, a pioneer in the study of quasars:

> *Horrid quasar*
> *Near or far*
> *This truth to you I must confess;*
> *My heart for you is full of hate.*
> *O super star,*
> *Imploded gas,*
> *Exploded trash,*
> *You glowing speck upon a plate,*
> *Of Einstein's world you've made a mess.*

And consider this poem, found on the body of a particle physicist who committed suicide after hearing a talk on quasars:

> *Quasar, quasar shining bright*
> *Both by radio and by light*
> *Is it true that gravity*
> *Is thy source of energy?*
> *Quantum theory may break down*
> *Under pressure like your own*
> *Leaving theorists to say*
> *Quasar, won't you go away?*

Despite working nearly every day in an observatory, I still find mind-boggling our burgeoning cosmic awareness that we can be conscious of objects so far and foreign. Though relegated to a minute

21

cranny of the Universe, we have recently extended our cosmic inventory to realms undreamed of mere decades ago. We may worry about our inability to explain quasars, but in my opinion, just knowing of their existence is a splendid accomplishment.

Some years ago I used MIT's big Haystack radio telescope near my hometown in northeastern Massachusetts to observe the most distant object then known in the Universe. Called only by its catalog name, OQ172, this quasar is nearly thirteen billion light-years from Earth. In English units, that's some 80,000,000,000,-000,000,000,000 miles away. Regardless of the units used, this cosmic candle is terribly remote. But it's not the concept of distance that astounds me. It's the concept of time.

Working the equipment in the control room, I was enthralled to realize that OQ172's radio signals had been traveling through the near void of outer space for some thirteen billion years. To witness its radiation tripping the receiver and rousing the recorder was mind-expanding, a revelation of sorts. My thoughts drifted into the past, the truly distant past. The radio signals I examined were launched eons ago, in the earliest epochs of the Universe. More than midway in their cruise this way, the Sun and Earth formed; later life arose on our planet, then intelligence, civilization, and technology. Still, the signals raced onward at the speed of light. With the quasar's radio signals nearing Earth, I was born, my radio telescope was built, and together we met one night a few years ago. I had turned my telescope into a time machine. Though I had previously disdained historians, I had become one. Indeed, astronomers are the real historians; we truly watch the past unfold.

Observing the cosmos creates an eerie feeling, an uncanny experience. While most citizens sleep, astronomers are one with the Universe, straining to fathom, yearning to understand, mentally plugged into the grand scale of worlds far, far away. The remote stars and galaxies broadcast information just as surely as the television sets in our living rooms, not about economics, politics, or sociology but about our cosmic roots. I'm convinced that if social, political, and religious leaders were to observe the cosmos and to grasp its big picture, they might then realize how we fit into the cosmic scheme of things, how fragile yet beautiful our spaceship Earth really is. In doing so, we might just get our earthly act together for the betterment of all humankind.

So, a zoo of galaxylike objects litters our Universe. But astronomers seek to know more. We are currently struggling to understand how these grand objects originated and especially how they change over the course of time. What evolutionary schemes link the various types of galaxies strewn throughout the Universe? Much akin to the nature-nurture struggle in the living world, do galaxies change intrinsically (read "genetically") or are they influenced largely by their intergalactic environments (read "culturally")? More generally, does matter transform inherently or in response to some external driving force?

The answers are unknown. But here are the broad outlines of an elegant idea now being tested at leading observatories around the world: A time sequence starting with quasars, proceeding then to active galaxies and finally to normal galaxies suggests a continuous range of cosmic energies. For example, weak quasars have some properties akin to those of the most explosive types of active galaxies, while the feeblest active galaxies often resemble the most energetic members of the normal galaxies. Such a chain of cosmic verve implies that all galaxylike objects might have initially formed as quasars roughly ten billion or more years ago, after which their emissive powers gradually declined. Over the course of time the quasars evolved into active galaxies and eventually into normal galaxies.

Why do we see any quasars at all, many billions of years after the bang? Because radiation needs many billions of years to reach us from so far away in space. We see the remote quasars as they once were in their blazing youth, not as they are now. Should this idea be correct, then the quasars are the ancestors of all the galaxies. Even our Milky Way might have once been a brilliant quasar. Ironic, indeed: Humans gape in awe at the herculean quasars, yet we might well live inside a time-tamed version of one. Perhaps.

If our knowledge of galaxies seems a little fuzzy, that's because it *is* a bit fuzzy. Our current grasp of the origin and evolution of galaxies is clearly inadequate. How the many varied galaxies came to be, endowed with their peculiar shapes and prodigious energies, remains one of the great missing links in all modern science. The riddle is deep because galaxies themselves provide few clues—at least clues yet recognizable to humans. Scanning the skies far and wide, we cannot find any galaxy in the act of formation. Nor have we

23

detected any unambiguous evolutionary changes among the galaxies.

Given the short duration that our technological civilization has thus far probed the cosmos, I'm not surprised that galaxies seem immutable. Earthlings emerged very recently in cosmic time, we evolve in a flash compared to the rest of the Universe, and we endure for a mere wink in the cosmic scheme of things. Who are we to fathom nature's grandest structures? Then again, I often find myself musing that galaxies are nothing more than huge collections of atoms. Cerebral matter in our human skulls is vastly more complex than a bunch of galactic atoms, however huge. We should be able to decode the galaxies. Indeed, we must if we are to appreciate the full grandeur of cosmic evolution.

> I believe a leaf of grass is no
> less than the journey-work of the stars.
> —Walt Whitman

Stars are the clearest evidence of the Universe beyond. Sized midway between the smallest and largest of all known objects, these hot balls of gas are larger than atoms by roughly the same factor of a billion billion by which they are dwarfed by galaxy clusters. Of median size, yes, but stars are not of middling significance. From the viewpoint of living beings, stars are practical necessities for at least two reasons.

First, stars are the furnaces that forge heavy matter. Colliding viciously, light nuclei of hydrogen and helium fuse into the ninety-odd chemical elements, such as carbon, oxygen, nitrogen, silicon, and iron. Without the heavies, nothing around us—not the ground, not the air, not even leaves of grass—would exist. Second, stars play essential roles in energizing nearby planets. At home our Sun helped brew the chemical stew that led to life on Earth. Emitting more energy *per second* than humans have generated in all of history—the equivalent of about a billion one-megaton nuclear bombs—our Sun pours forth the heat and light needed for the further development and continued maintenance of life on our planet.

Indeed, stars are prerequisites for life. More than anything else, solar energy converts the rock called Earth into a reasonably comfortable abode. Without a nearby star, Earth would be a frozen,

24

barren wasteland. At 270 degrees below zero Celsuis, it would resemble a boulder so unimaginably hostile that life as we know it
could not possibly exist.

Gazing at the nighttime sky as a boy, I found stars boring. A
little mystical, but still boring. Strange statement by a future astrophysicist? Perhaps. But to the uninformed, stars are terribly unexcit-
ing in their seemingly unflinching ways. Night after night, year
after year, stars hover over us, appearing unchanged. No wonder the
ancients took them to be fixtures on a mammoth celestial dome—a
fresco on an astronomical ceiling. I now know better. Stars appear
immutable only to our naked eyes and only from our human perspective.

During the past few decades observations have proved that
some stars are old, some young. Many are long gone, having literally
run out of fuel and died eons ago. Still others are only emerging from
the galactic mishmash of gas and dust. We now know a great deal
about how stars are born, progress, and die—how they experience
phases of youth, maturity, and aging. Giant stars such as reddish
Betelgeuse in the constellation Orion, dwarf stars such as the
whitish companion to Sirius the Dog Star, and average stars such as
our yellowish Sun are not really different types of stars. Rather, each
is at a different stage in the changing life cycle of nearly all stars.

Nothing more basic confronts astronomy than the birth of
stars. Yet until this past decade we were unsure how stars form.
Theories were plenty, but observations were lacking. The problem is
that stars are conceived in dense, dark, and dusty galactic clouds
where quite frankly, there is nothing to see; not unlike the mammals, the brightest stars are incubated in total darkness. But don't
let the darkness fool you. Having peculiar names like Rho
Ophiuchus, Messier 20 Southwest, and R Monoceros, the darkened
tracts amidst the nighttime stars harbor a great deal more than
nothing. Indeed, the gas and dust in galactic clouds are the stuff of
which stars are made.

Throughout portions of interstellar space, galactic gas surges,
accumulates, contracts, and warms. All the while, the dark clouds
radiate invisible radio and infrared waves. Even at this moment
antennas and telescopes around the world are monitoring the galactic darkness above our heads. There, once their placental envelopes
have broken, compact blobs occasionally emerge as stellar bodies—

bright and shiny balls of gas. Looking blue-white, cool and icy, newborn stars nonetheless boil at temperatures far greater than those of the hottest steel furnaces built by human civilization. Tens of thousands of degrees Celsius at their surfaces and tens of millions of degrees at their cores, the young stars clustered near the center of the Orion Nebula are good examples of recent star formation. The nebula itself, a glowing cloud of plasma light-years across, is a cosmic version of a neon sign flashing "Birthplace of Stars."

Hard to imagine, but the pinpricks of light in the evening heavens owe their existence to nuclear fires churning deep in the cores of each and every one of them. Contemplate all that astronomical activity, all those cosmic nuclear reactors in the sky. Go outside some clear, moonless evening, gaze upward, and just ponder all that hidden effervescence.

The remarkable change from galactic cloud to contracting blob to nascent star spans a few tens of millions of years. Several million generations by human standards, this is still less than one percent of a typical star's lifetime. The entire process amounts to a steady metamorphosis, an evolution, a gradual transformation of a cold, tenuous, flimsy pocket of gas into a hot, dense, round star.

Once the outward-pushing heat of the nuclear inferno balances the relentless onslaught of inward-pulling gravity, equilibrium is achieved. A fierce struggle ensues between heat and gravity for billions of years. So long as this cosmic tug-of-war remains a stalemate, dramatic changes are virtually nil. But no star can shine forever. Eventually all stars perish, some more catastrophically than others. For example, our Sun and all other average stars like it will creep toward oblivion while running out of fuel. Swelling into a red-giant star, our bloated and aged Sun will someday engulf the inner planets Mercury and Venus. Earth's surface will roast, its oceans evaporate, its atmosphere dissipate. No need to panic; if our calculations are right, this hell-on-Earth will not commence for another five billion years.

Thereafter our Sun and all stars like it are destined to become planetary nebulae much like the Ring Nebula in Lyra. Outer space is populated with dozens of these weird-looking objects having flimsy, tenuous halos surrounding dense, Earth-sized cores—inner lumps so dense that a thimbleful of their matter would weigh a ton on Earth. Called white-dwarf stars, such compact cores gradually cool and shrink, becoming yellow dwarfs, then red dwarfs, and finally black dwarfs—dark, dead clinkers in space. Though we are currently unsure, the darkened regions of our Galaxy might well be strewn with many such stellar corpses already consigned to the graveyard of stars. In truth, galactic space could be chock-full of interstellar basketballs, and we'd have no way of knowing.

Let me stress again that although change is the hallmark of stellar evolution, that change is woefully slow. Most of it transpires over time scales much longer than even the ten-thousand-year duration of our civilization. So no humans—not even astronomers—can watch individual cosmic objects march through their full pageant of stellar evolution, from emergence in dust to thrust toward doom.

Fortunately, infrared telescopes, such as NASA's versatile mirror on Mauna Kea, Hawaii, and radio telescopes, such as the big Arecibo antenna in Puerto Rico, are now probing galactic clouds and nascent stars for hints about their embryonic development. Likewise, Earth-orbiting satellites such as the *Copernicus* Ultraviolet Observatory and the *Einstein* X-ray Observatory have permitted us to examine aged stars and stellar remnants for clues to the fates of celestial bodies.

In the course of this work astrophysicists resemble anthropologists. Not having lived at the time of our ancient ancestors, anthropologists sift through earthly rubble, retrieving artifacts and bony remnants from unrelated localities scattered across our planet. Their objective, of course, is to decipher how the remains can be joined together to yield an overall picture of human evolution. Likewise, astrophysicists observe various objects in unrelated regions of our Galaxy. Finding some young objects here, some old ones there, and more than a few that boggle our minds, we try to diagnose how each object fits into the overall scheme of stellar evolution. Much as in jigsaw puzzles, terrestrial bones and extraterrestrial objects are the pieces. The picture becomes clear only when each fragment is found, identified, and fitted properly into its place among all the other pieces.

Different fates await stars much larger than our Sun. Once the nuclear fires cease, gravity overwhelms all forces, causing a massive star to collapse. In a matter of seconds the imploding sphere detonates its core, rebounds like a coiled spring, and ejects its surrounding layers. Much of its mass, including a host of heavy elements cooked within, is expelled outward. The result is a shattered star strewn chaotically throughout neighboring regions of space.

Such signposts of stellar death are known as supernovae, the most famous of which is the Crab Nebula, a remnant of scattered debris some five thousand light-years distant in the constellation Taurus. Observed and reported in A.D. 1054 by Arab and Chinese astronomers (as well as apparently by some tribes of midwestern American Indians), the glowing gases of the Crab's death rattle once rivaled the brightness of our Moon. The exploded star reportedly could be seen in broad daylight.

Given the number of stars in our Galaxy, we can expect a supernova to burst forth every century or so. However, a viewable star in our Galaxy has not blown up in this way since Galileo Galilei first used a telescope early in the seventeenth century. Hence, our Milky Way seems long overdue for a supernova. Who knows? Any day now we may be treated to nature's most spectacular exhibition.

Actually, supernovae might be more than splendid light shows. Should a massive star detonate in the galactic suburbs near our home, it could well inundate Earth with radiation harmful to life. At least one theory suggests that a supernova contributed to the sudden extinction of the dinosaurs. Based on a recent stellar census, our best estimates suggest that a supernova should occur within roughly thirty light-years of our Sun once every half billion years. Too close for comfort? Fortunately, none of our neighboring stars is currently massive enough to die catastrophically by exploding in this way. Luckily for us, all our immediate cosmic neighbors seem destined to perish, as will our Sun, via the more placid red giant-white dwarf route.

I'm fascinated by the notion that one or more of the stars we see in the nighttime sky have almost surely *already* exploded, but the light from this stupendous event has yet to reach our planet. Some of the bright stars could have blown up decades or even centuries ago, and we wouldn't yet know it. The massive star Rigel, for instance, looking proud and mighty some nine hundred light-years from Earth, could have exploded during Earth's Dark Ages. Starry messages take time to race across even the relatively nearby heavens.

What remains in the aftermath of a supernova explosion? Is the entire star just blown to bits into the interstellar void? We're not quite sure, though most theoretical models predict that some part of the star survives. Like planetary nebulae that leave behind cindral reminders of stars' former brilliance, we expect supernovae to bequeath remnant cores. Called neutron stars, supernova cores have been found amidst the debris of previously exploded stars. For example, more than a decade ago astronomers discovered the core of the Crab Nebula to be pulsating nearly thirty times per second. The culprit turned out to be a classical "pulsar," another name for a rapidly rotating neutron star.

A neutron star represents one of the strangest states of matter in all the Universe. Not really a "star" at all, it's rather an ultradense ball of subatomic particles compressed to a size not much larger than a typical city. There gravity is so intense that a teaspoon of neutron-star stuff would weigh on Earth about a million tons, a human on its surface would be crushed to the thickness of a postage stamp, and the entire population of planet Earth, if shipped to a neutron star, would be squeezed into a volume the size of a pea! Astoundingly, if a single Egyptian pyramid were made of matter this dense, its gravitational attraction would be sufficient to destroy the Earth. Strange objects, these neutron stars.

As best we can tell, neutron stars are stable, but not because of the usual gravity-in, pressure-out conflict. Instead, the outward force arises from the crystalline nature of the supercramped neutrons. Virtually touching one another, the neutrons form a solidified ball of matter that not even gravity can compress further—with one notable exception.

According to some theories, if three or more times the mass of the Sun are jammed into an extremely small space, then gravity can overwhelm any countervailing force. And without anything to stop it, gravity alone will steadily compress a massive star to the size of a planet, a city, a pinhead, a microbe, even smaller! Collapsing into minute regions, such senile though massive stars apparently disappear, trapped forever by their own gravities. Incommunicado, for nothing can escape the gravitational pull of such regions, these bizarre end points of stellar evolution are termed black holes. Not objects, mind you, but "holes," and ones that are dark to boot.

Until recently I regarded black holes as convenient cop-outs for unexplained cosmic phenomena. It seemed that every time astronomers discovered a new type of object during the last decade, someone claimed it as a kind of black hole. I remember chuckling several years ago when Britain's premier astrophysicist, Martin Rees, argued at a Boston symposium that gigantic black holes lurk in the hearts of most galaxies. What a vivid imagination, I thought! But I've recently changed my tune, for observations—the acid test of any theory—have now strengthened the case for massive black holes.

For several years I actively participated in the exploration of the central regions of our Milky Way Galaxy, some thirty thousand light-years from Earth in the direction of the constellation Sagittarius. Using radio and infrared techniques, a few of us helped discover there a fiercely swirling vortex of ten-thousand-degree gas spread over some ten light-years. As if this whirlpool were not peculiar enough, our observations imply that deep in its midst lies a heretofore unrecognized and compact "something" several *million* times more massive than our Sun. Gulping perhaps whole stars, and presumably growing, this bizarre supermassive region is probably a black hole. Rees might well be right after all.

Throughout our research we use extra caution, for we are still learning to grope in the dark, to sift through the clues hidden within invisible radiation. Comprising nothing less than a galactic ecosystem, the evolutionary balance among the many varied stellar and interstellar components might be as complex and delicate as that of life in a tide pool or a tropical forest. Only by being receptive to information from the Milky Way in all the electromagnetic frequencies at which it chooses to radiate can we hope to understand some of nature's cycles within our Galaxy. Accordingly, I often regard myself as more an explorer than a scientist. Even anthropologists can see and feel what they study; when it comes to black holes, astronomers can neither see nor touch.

What's it like deep down inside a black hole? No one really knows. Perhaps the question itself is irrelevant. Though experiments to test a hole's nature might well someday be mounted on robot space probes sent "down under," that information could never reach the rest of us outside. Nothing—not even light—can escape a black hole.

Here's an existentialist's interpretation by the Frenchman Jean-Paul Sartre:

. . . I enter the black hole. Seeing the shadow at my feet lose itself in the darkness, I have the impression of plunging into icy water. Before me, at the very end, through the layers of black, I can make out a pinkish pallor. . . . I stop to listen. I am cold, my ears hurt, they must be all red, but I no longer feel myself: I am won over by the purity surrounding me; nothing is alive, the wind whistles, the straight lines flee in the night. . . .

To others, such as Mark Twain, black holes apparently need not be black, only "a totally new kind of hole," "a perfectly elegant hole," or "a long hole and a deep hole and a mighty singular hole altogether." To William Butler Yeats:

> Things fall apart; the center cannot hold,
> Mere anarchy is loosed upon the world.

And to e. e. cummings:

> unwish through curving wherewhen
> til unwish returns on it unself.

As a native New Englander I prefer the earthy Robert Frost, who, from yet another humanistic perspective, long ago predicted the fate of our galactic center:

> We dance round in a ring and suppose,
> But the Secret sits in the middle and knows.

Whatever, the Universe did emerge some fifteen billion years ago from what would seem to have been a "naked singularity" resembling a black hole. Such baffling singularities, like the X-1 source in the constellation Cygnus (the Swan), could well be the key needed to unlock an understanding of the creation state from which the Universe arose. By theoretically studying the nature of black holes, especially by experimentally seeking their existence and physical properties, we may someday be in a better position to appreciate *the* most fundamental issue of all—the origin of the Universe itself.

31

Worlds on worlds are rolling ever
From creation to decay,
Like the bubbles on a river
Sparkling, bursting, borne away.
—Percy Bysshe Shelley

Planets are globes of solids, liquids, and gases, smaller than stars and made partly of heavy elements. We know these worlds must have formed in relatively recent times, for heavy elements did not exist early in the Universe. Planets had to await the birth and death of countless massive stars, the cores of which bake the heavy elements. Planets, then, are heaps of cinders from burned-out stars, swirls of ashen dust created during stellar blasts.

I often regard this galactic modus operandi as a kind of cosmic reincarnation. The act of stellar death ensures a recurring fertilization of interstellar space, out of which emerge in turn later-generation stars and planets. In time the stars' elemental ashes provide the seedlings for life itself. On and on the cycle churns. Stellar buildup, breakdown, change. Dust to dust, and to dust some more.

Our planetary swarm is a varied lot. Called the Solar System, we now know it includes one star, nine planets, about three dozen moons, thousands of asteroids ranging in size up to a few hundred kilometers, myriad comets of kilometer dimensions, and innumerable meteoroids less than a meter across. With the Earth–Sun distance of about a hundred million kilometers termed an "astronomical unit," the whole Solar System extends end to end for nearly eighty such units. That may sound large, but it's only about a thousandth of a light-year, hardly a billionth the size of our Milky Way—a few blocks by cosmic standards.

From the perspective of, say, the nearby Alpha Centauri star system, Old Sol overwhelmingly dominates our neighborhood, with Jupiter an inferior second. In fact, our star houses more than 99.9 percent of all the matter in the Solar System. Everything else amounts to wads of nearly insignificant debris. Astronomically, Earth is just one such piece of scrap.

But be sure to distinguish Jupiter from the other planets, for this striped, airy ball is no ordinary heavenly body. Jupiter just missed becoming a star. Unlike Earth but like our Sun, Jupiter harbors vast quantities of hydrogen and helium gas. Vast, to be sure,

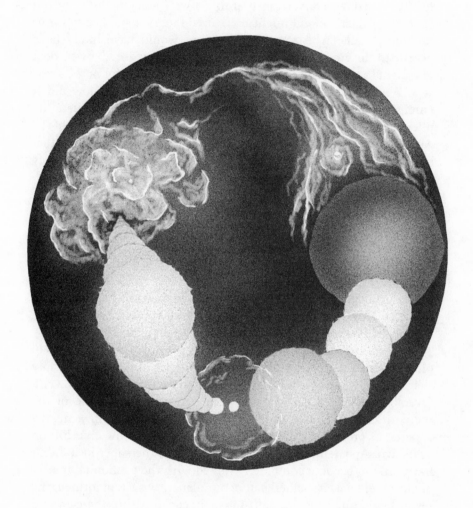

On and on the cycle churns. Stellar buildup, breakdown, change. Dust to dust, and to dust some more . . . a kind of cosmic reincarnation.

33

but not quite large enough to self-ignite. Had Jupiter amassed a few tens of times more matter, its central gases would have commenced nuclear burning, converting it into a dwarf star. In this way our ancestral galactic cloud might have sired binary stars. Lucky for us it didn't; such an astronomical posture would have likely been hostile to the emergence of life on Earth. So I suppose we owe a debt of gratitude to the Sun for lighting up, and to Jupiter for not.

Whence did our Solar System come? When and how did it emerge from the chaotic hodgepodge of galactic gas and dust? Replies to these questions depend upon models. Much like those of the early Universe, our models are computer-crunching exercises that seek to explain the origin and architecture of the Sun and its family of planets and moons. Loads of facts embellish the models, including data acquired from our studies of interstellar clouds, interplanetary meteorites, as well as Earth's Moon.

Especially helpful is the wealth of information now being telemetered to Earth by American and Soviet robot space probes reconnoitering the nearby planets. The U.S. *Voyager* spacecraft that bypassed Jupiter and Saturn several years ago yielded particularly striking insights into planetary meteorology, structure, and composition. Now racing out of the Solar System, *Voyager 2* encountered Uranus in 1986 and is scheduled to visit Neptune in 1989.

Meteorite studies have proved even more fruitful. Analyses of old, uneroded meteorites recently found on the Antarctic ice imply that the creation of our Solar System was a rather violent event. Overabundances of carbon, nitrogen, and oxygen in some of the meteorites' dust grains suggest that a concussion from a nearby supernova probably triggered planetary genesis some five billion years ago. Apparently the ejected debris from the supernova didn't have time to become completely mixed with the primordial matter of our parent galactic cloud before our planetary system formed, the result being microscopic inclusions embedded within the captured meteorites.

Nuclear weapons tests conducted at the Pacific's Bikini Atoll during the early 1960s demonstrate what probably happened next. Just as an atomic bomb blast implodes houses (rather than explodes them), a supernova's shock wave would have raced around the thin exterior of our ancestral galactic cloud, squeezing it from many directions and compressing it into a tangible blob. The shock itself,

34

. . . a concussion from a nearby supernova probably triggered planetary genesis some five billion years ago.

so say our models, naturally produces whirling eddies of gas at various sites strewn about a rotating, Frisbee-shaped disk of primordial matter. You can witness a similar phenomenon by watching eddies form behind a sailboat cruising a lake or by stirring a spoon through a cup of tea. Once an eddy gathers enough matter while orbiting the protosun, then gravity alone ensures the eventual formation of a planet. I sometimes liken the accretion process to a snowball thrown through a fierce winter storm; the ball fattens by sweeping up more snowflakes in its path.

Provided the "sweeping" was reasonably efficient throughout the primordial disk, we can appreciate how our present Solar System has come to exist as a collection of rather tiny planets wheeling around a huge sunny sphere in an otherwise empty region of space. Our models suggest that about a million centuries were needed to generate nine protoplanetary eddies, scores of protomoons, as well as a big protosolar eddy in their midst. Roughly a billion more years were needed to sweep the system reasonably clear of interplanetary trash.

Present-day comets, such as Halley's recent apparition, are vestiges of this erstwhile formative stage. So are meteorite showers, such as the Perseid Meteors that rain down each August. Comets and meteorites serve as reminders of birth and construction, not as omens of death and destruction.

A decade ago I had the impression that we inhabited a cosmic junkyard. Scattered remains of peculiar planets and desolate moons told of a violent past. I thought it a hopeless task to unravel, seemingly impossible to be sure of the forces that once shaped our earthly abode. But the outlook is now changing. Buoyed largely by the startling success of NASA's program of planetary exploration, we've been forced to revise our perceptions of these rambling worlds. The revision is a healthy one, for we are only now beginning to recognize the puzzle pieces of a frenzied youth, but a richly constructive one as well.

In much the same way that red giants and white dwarfs represent varied stages of stellar evolution, many planets and moons now seem to be at different levels of the planetary life cycle. For example, the big Jovian Planets—Jupiter, Saturn, Uranus, and Neptune—are galactic fragments frozen in time, not massive enough to become stars yet too hot inside to fashion solid rocks. To varying degrees, these gassy worlds preserve the pristine properties of the primitive Solar System. Alternatively, the less massive Terrestrial Planets—Mercury, Venus, Earth, and Mars—have evolved a great deal, cooling and crystallizing hard, rocky surfaces, while exhaling atmospheres and sometimes exuding oceans. Significantly, at least one of these small planets has also spawned life.

So as we get to know the planets better, we should view the Solar System not just as a collection of planetary refuse. Every planet and moon have something to tell us, something about their origin and evolution. Each time a new space probe closely encounters a planet, we learn a little more about that alien world. The Russian *Venera* crash landings on hellish Venus; the American *Viking* robots parked on the dusty plains of ruddy Mars; the *Voyager* mission now skirting the Jovian Planets and their myriad moons: With these and other artificially intelligent robots, we can feel some of the excitement akin to Europe's Renaissance, as their explorers probed the New World of the Americas.

We live at a particularly auspicious time in human history—a golden age of cosmic exploration, a second Renaissance of sorts. If the rapid rate of our discoveries continues in the 1980s and 1990s, I'd guess that a blend of all planetary studies will soon yield a thorough grasp of the origin, evolution, and destiny of our island home within this vast Universe. Then, and only then, will we have framed our worldly heritage.

> *There's nothing constant in the universe,*
> *All ebb and flow, and every shape that's born*
> *Bears in its womb the seeds of change.*
> —Ovid, "Metamorphoses," XV

Planet Earth is peppered with heavy elements fused in supernova death throes. We need not be clever chemists to realize that the air, land, and sea are partially made of heavies. Nitrogen in the air, silicon and iron in the rocks, oxygen in the water are among many examples. Life itself is the most remarkable of all heavy-element concoctions. Indeed, stars have died so that we might live.

Life everywhere now seems biologically adapted to our planet. But adaptation is a never-ending effort. Change is inevitable. Nothing is immutable, nothing at all. The climate changes. The

Alps build. The Atlantic widens. Even the rock-solid aspects of sturdy Earth quake and drift, evolving over time scales immense compared to human life spans. What we can't see is tough to believe, but we are witness to so brief a time. Even the ten-thousand-year duration of human civilization is a mere wink in the spectacle of fifteen billion years of cosmic change.

Consider the host of plants and animals on our planet. Redwoods and rhinos, roses and reindeer, evergreens and elephants, uncounted more species. Where did they all come from? That they suddenly appeared intact from nothing is an interesting idea. But spontaneous creation makes no sense scientifically, nor is there a shred of evidence to support it.

Biochemists who study molecules and cells are now pooling their talents with paleontologists who study fossilized bones. Together they are gaining a rich understanding of how life emerged and evolved eons ago. To appreciate their efforts, we must imagine the early Earth some five billion years ago.

Shortly after Earth formed, it was hot, oceanless, lacking in oxygen, and pelted with all sorts of energy from within and without. Solar ultraviolet radiation, intense meteoritic bombardment, fierce thunder and lightning, radioactive rocks, and violent volcanoes all energized our young planet. Gradually, restless Earth cooled, cracked, exhaled steam, and outgassed an atmosphere. Much of the water vapor thereafter condensed, secreting the oceans.

Using flasks, test tubes, and a lot of rubber hosing, we now know how to build laboratory apparatae to simulate Earth's early ocean and atmosphere. These chemistry experiments show how gaseous hydrogen, H, carbon, C, nitrogen, N, and oxygen, O, collide, stick, and react; these gases would have done so naturally on primitive Earth. The result is a mix of simple molecules, including ammonia, NH_3, methane, CH_4, carbon dioxide, CO_2, and water vapor, H_2O.

Further tests show that these molecules have a tendency to combine into more complex chemicals, provided they are gently "cooked" with some energy or radiation. Here energy plays the role of catalyst, helping rupture some bonds of the small molecules, while guiding the formation of even larger molecules. Actually, energy is more than a catalyst. Energy fashions a near miracle: It helps produce the precise molecular building blocks of life. These

38

. . . gently "cooked" with some energy or radiation.

are the amino acids that make proteins such as hemoglobin and insulin and the nucleotide bases that comprise genes such as DNA.

I find it ironic that people seem frightened these days by mention of the words "chemicals" and "radiation." These are the foundations of all life. To paraphrase an American television commercial, "Without chemicals and radiation, life itself would be impossible."

Though I regard these chemistry experiments as terribly significant, I sometimes confess confusion. Life around me seems so complex, so varied. How could the essence of life be made so easily in a test tube? The answer, I must remind myself repeatedly, is that life is not overly complex physically and chemically. Humans may have made it so, culturally and socially. But when it is reduced to its component parts, the basic ingredients of life—any life, from bacteria to ostriches to humans—are hardly more exotic than two dozen molecules. And these molecules are the same forerunners of the genetic and proteinlike substances produced in the laboratory experiments.

Genes and proteins: The first directs reproduction, the passage of heredity from one generation of life to another; the second directs the metabolism, the daily flow of incoming energy (food) and outgoing wastes. In man, mouse, or daisy, the genes mastermind life, and the proteins maintain its well-being.

Be assured, nothing living crawls out of the primordial soup in the laboratory experiments. Not even simple cells have yet been made under test-tube conditions. But the commonality of the molecules comprising *all* life on Earth—the very same molecules synthesized in the lab—is our best evidence that every living thing dates back to a single ancestor billions of years ago.

How do we know anything about the previous episodes of life on Earth? What information do we have about life forms long ago dead and buried? We are guided by many clues preserved in the rocks of our planet. Most of these hints are fossils—visible traces of dead organisms that once lived. I can recall, for example, especially while digging in our nation's Southwest, discovering fossils of weird and unfamiliar species. That peculiar creatures like trilobites and dinosaurs once roamed our planet at first staggered my mind. But we now realize that the rocks themselves embody an unmistakable trend: Old rocks embed only simple life, whereas young rocks contain mostly complex life. The bizarre fossils are invariably found within older rocks. The youngest of all rocks house fossils of familiar life forms—chimps, horses, men and women, among others.

Together all the fossils chronicle an amazing story of life on Earth. Repeatedly, throughout the millennia, new life forms emerged, while others perished. Some species survived for ages; others succumbed as soon as they appeared. Incredibly, more than ninety-nine percent of all life forms that once prospered are now extinct.

Beginning nearly four billion years ago, life progressed erratically through the ages. For the first few billion years it remained simple, resembling the blue-green algae now found in backyard swimming pools. Eventually, around a billion years ago, life forms clustered, coordinated their activities, and became multicellular organisms much like the (real) sponges now used in our bathtubs. Shortly thereafter the fossil record documents what must have been a population explosion in the number and diversity of species.

Change was rampant. Legions of fish populated the seas about six hundred million years ago. Plants came ashore some four hundred million years ago. Amphibious fish quickly followed, doubtless in search of food. Animals mastered the land about two hundred million years ago, while birds, mammals, and flowers flourished nearly a hundred million years ago. And men and women? Well, hominids have endured for only the past few million years, a brief span indeed in the entire scenario of cosmic evolution. In fact, if we imagine the whole history of the Universe compressed into a single year, then earthly humans have existed for only the past hour or so. In this analogy our specific species, *Homo sapiens*, did not emerge until some ten minutes ago.

Charles Darwin was basically right. The fossil record leaves little doubt that biological evolution by natural selection has occurred and is continuing. It is as much a fact as the Earth's going around the Sun. The rate at which evolution works, however, remains unsolved and controversial. In brief, we now recognize that chance mutations act as the motor of evolution. Caused by cosmic rays, chemical drugs, or intense radiation, mutations alter life's genetic structure, enabling some organisms to achieve the best available niches within ever-changing environments. And I'll stress again that changes, and adaptations to those changes, are the keys to the genesis—and survival—of all things.

41

Beginning nearly four billion years ago, life progressed erratically through the ages.

Of course, some gaps in the fossil record hinder our complete understanding of life's history, just as some missing links hamper our current knowledge of galaxies, stars, and planets. But each day brings new discoveries, new tests, and further refinement of our modern ideas of biological and cultural evolution. And with these advances come greater objectivity, and progress too, in our search to understand reality.

Tree swinging, manual dexterity, binocular vision, fire, tools, speech, writing, foresight, curiosity: These are among the evolutionary developments that helped make us human. They had a clear effect on the brain: It got bigger. Doubtlessly the gray cells within our skulls are the most extraordinary clumps of universal matter known. As best we can tell, the human brain is the ultimate example of complexity in the physical Universe.

I regard the brain as a living machine, a remarkable cluster of star stuff, one that enables us to create an inquiring civilization, to unlock secrets of the Universe, and to reflect upon the material contents from which we arose. Indeed, with the onset of technological intelligence on our planet and perhaps on others as well, sentient beings learn to manipulate and dominate matter, much as matter evolved earlier to govern energy. *This change, from matter-dominance to life-dominance, I claim is the second of two preeminent events in the history of the Universe.*

Without a brainy seat of consciousness and its inherent awareness of self and environment, galaxies would twirl and stars would shine, but no one or thing could comprehend the majesty of the reality that is nature. With a brain, we probe the past, striving to decipher our celestial roots, all the while searching for a better understanding of the cosmos and especially of our true selves. What emerges is nothing less than a cosmic heritage, a plenary view of who we are, where we came from, and how we fit into the universal scheme of all things material.

> *Though leaves are many, the root is one;*
> *Through all the lying days of my youth*
> *I swayed my leaves and flowers in the sun;*
> *Now I may wither into the truth.*
> *—William Butler Yeats*

. . . galaxies would twirl and stars would shine, but no one or thing could comprehend the majesty of the reality that is nature.

44

I suggest that cosmic evolution is a powerful synthesis to use as perspective—a grand ethos of potentially unprecedented intellectual magnitude—while approaching an uncertain future. Looking backward, we sense that its central feature—the time-honored concept of change—can account for the appearance of matter from the primal energy of the Universe and in turn for the emergence of life from that matter. Change further seems capable of describing the act of creation itself, thus scientifically accounting for the origin of all energy at the alpha point of space and time.

But are we Earthlings to survive to learn more about ourselves, our planet, our Universe? Looking forward, shall we achieve some astronomical destiny? Just how wise, quite aside from sheer intelligence, are we? Put bluntly and not insignificantly: From the study of cosmic evolution may well emerge a sense of "big thinking" and with it the global ethics and planetary citizenship needed if our species is to have a future. In the words of Sören Kierkegaard, "Life can only be understood backwards, but it must be lived forwards." Tritely stated, though no less true: Our future will likely be a measure of our current wisdom.

CHAPTER 2

History of
Change

THE IMPORTANCE OF THE CONCEPT OF CHANGE
HAS ITSELF CHANGED THROUGHOUT HISTORY

THE CONCEPT OF CHANGE is not new. As noted in the previous chapter, many basic features implicit in our modern idea of evolution date back at least to the time of the early Greeks some two dozen centuries ago. And by "evolution," I mean here the developmental change of all things, not just life forms and not solely by means of Darwin's major explanatory principle of natural selection.

At the outset let me stress that this chapter is not meant to be a synopsis of the history of philosophy, or even of the history of the philosophy of science, as much as of the history of the (largely) Western concept of change. As such, some of humanity's most influential scholars, including, for example, the ancients Socrates and Plato, relative moderns Galileo and Hegel, as well as near contemporaries Freud and Bohr, will nary rate mention, for as best I can tell, they contributed little to the advancement of evolutionary insight. I limit my survey in this way knowing full well that a given feature—in this case, change—of a larger philosophical system is often difficult to appreciate by being removed from context.

I also acknowledge my disproportionate attention to early Oriental philosophy. Although the cryptic *I Ching* (or *Book of Changes*; c. 1200 B.C., including the allegedly Confucian appendices c. 500 B.C.) posits cyclic interchange of the passive yin and active yang (two forces begotten by the Great Ultimate), Eastern speculation regarding the concept of change seems not to have developed a historic richness as did philosophies of the West. Taoists of classical China apparently intellectualized mainly in the subjects of chemistry and morality, anticipating little of the evolutionary perspective or of the relationship of humans to all living things, while the orthodox Hindu text Upanishads (c. 800 B.C.) held the material cosmos to be an illusion, thus prompting Indians of antiquity to regard human inquiry into the nature of things as forever inconsequential. As has been suggested by a number of writers, the notion of evolution as progress depends crucially on the concept of linear time. Philosophies, religions, and other worldviews of the East are (even today) dominated by cyclical time and thus only grudgingly acknowledge the concept of progress. Indeed, as chauvinistic as it may sound, today's technological society developed primarily in the West largely because only we in the West historically possessed a linear conception of time.

By placing the study of the idea of evolution into historical perspective, I aim to demonstrate the continuity in the development of the idea throughout the ages. I contend that the influences of early upon later thought are a good deal stronger than previously recognized. Even Darwin owed more to the naturalists of antiquity and others of his predecessors than many of us have been willing to acknowledge.

Until little more than a century ago speculation outran fact. The development of evolutionism before Darwin, especially throughout the Middle Ages, often slackened and even regressed. Yet our current understanding of evolution was not reached by any quantum leap: neither the magnanimous achievement by Darwin and Wallace, who independently recognized the idea of natural selection as a means to explain the rich history and diversity of life on Earth, nor, as is now the case, the even more comprehensive cosmic evolutionary scenario that strives to synthesize the myriad changes among all energy, matter, and life in the Universe. Each of these developments is a product of history and of a changing sociocultural milieu.

Evolutionism has itself evolved over the ages, principally through the progressive development of secondary ideas connected with it. Indeed, our conception of change in nature continues to change today, incorporating novel theoretical nuances as well as a steady stream of often subtle observational findings. I would venture to say that if we take a large enough temporal perspective, evolutionary thought throughout recorded history has been characterized by, if anything, continuity and gradualism. On a smaller time scale, however, say, on one of centuries, the history of such thought, perhaps like evolution itself, has been characterized by stasis punctuated by periods of rapid and discontinuous change. Thus, evolution is itself both a historical conception and a product of a certain historical development. And it is the history of the concept of evolution itself that provides an interesting case study in evolutionism.

$$3^2 + 4^2 = 5^2.$$
—*Pythagoras*

Among the greatest achievements of the ancient Greeks was their first conceiving of the world, the world as a whole—namely, the Universe—as an issue to question and to try to answer. Their inquiries were born not of some practical need but of the passion for knowing, which had seized humankind.

Nature had a considerable influence on the philosophies and historical systems of the classical thinkers. Contrary to popular opinion, some of their earliest known models of the origin and causes of the Universe were not wholly imaginative; rather, their ideas sprang a priori, to a certain extent, from observations. Nonetheless, they rarely paused to test their theories by means of further observations and experiments, and so did not suffer the disappointment and delays that often plague our modern attempts to unlock the secrets of nature.

Combining a limited sense of observation, a wide range of ideas, considerable independence of thought, and a tendency to generalize, the Greeks possessed genuine gifts of scientific deduction. These qualities, coupled with a lack of a priestly class, enabled them to accumulate "truth" by means of a sort of inspiration.

Ancient Greek philosophy of nature matured gradually and in three phases. Aside from the largely prehistoric view which held the mythological notion that gods everywhere permitted humans and monstrous beings to grow, like plants, directly from Homer's Earth, the first or pre-Socratic phase stressed initially pure philosophy and later a form of naturalism; it was during this latter period of naturalism that, for example, Heraclitus and Empedocles championed the idea of change but apparently failed to conceive of its slow stages of development. This phase gave way to a teleological view that postulated a Grand Designer, beginning with the theism of Socrates and the essentialism of Plato and thereafter made famous in the natural philosophy of Aristotle. Finally, during the fourth and third centuries before Christ, the Epicureans revived and fostered a naturalistic or materialistic conception of the Universe.

The earliest known attempts to substitute natural explanations for supernaturalism—the oldest record of replacing myth with critical thinking—were initially and largely based on a philosophical school of thought in many ways exemplified by the Ionian Thales (c. 625–547 B.C.). Especially intrigued by astronomy and the origin of things, Thales is perhaps best known for having predicted an eclipse of the Sun, a rather dramatic "change" from the viewpoint of Earthbound inhabitants. (Using high-speed computers to trace backward the positions of the Sun, Earth, and Moon, modern astronomers concur that a total solar eclipse did in fact occur in Asia Minor in the year 585 B.C.) Thales' predictive success might not have been entirely his own, however, as his hometown of Miletus had close cultural ties with Babylonia, whose "astropriests" had long known how to predict lunar and solar eclipses quite accurately. Doubtless, these commercial contacts with Babylonia (and perhaps with Egypt as well) provided the Greeks with rich cultural interaction which helped soften the prejudices and superstitions of earlier, mythological times.

Of importance for us here while we trace the history of the idea of change, Thales was apparently the first of a long line of philosophers to regard water as the original matter from which all things originate. Tradition has it that he preached, "Water is best."

This hypothesis is hardly surprising in view of the Greeks' commercial dependence on the Aegean Sea as well as their extensive familiarity with shore and ocean life. Nor is water's primacy an altogether foolish idea; as noted in the previous chapter, the theory of an aquatic or a marine origin of life is widely accepted in current scientific circles. Further, water is two-thirds hydrogen, and we do now know that hydrogen is the common ancestor of all the chemical elements.

Actually, Thales' idea is just one of a lengthy list of early Greek opinions that today are considered basically correct. Another such idea is his statement, astonishing for the time, that the Sun and stars are merely balls of fire. Even so, the Greeks knew none of the details that we have at hand in the late twentieth century—details that could be secured only by means of strict experimental inquiry, a method of acquiring knowledge disdained by the pre-Aristotelians who looked down upon all manual labor. We might say that the early Greeks were broadly correct in many of their thoughts, especially those based partly on observation, but were quite stymied in thinking that those thoughts (and observations) need not be tested. Reason alone will never make the unknown known.

Contemporary with Thales in the Milesian school of philosophy was Anaximander (c. 610–546 B.C.). In some ways a vague forerunner of Kant in cosmology and of Darwin in biology, Anaximander was in other respects imbued with much of the old mythology. Unquestionably influenced by Thales, his teacher, as well as by his proximity to the sea, Anaximander also stressed the importance of water, though he did not take it to be the single primal substance from which all things arise.

In what fragments remain of his philosophy, Anaximander apparently never did specify the nature of that primal undifferentiated matter, though he did endow it with rather ambiguous adjectives such as "unfamiliar" and "infinite," the latter not in the mathematical sense but rather denoting limitlessness. "It [the unspecified primordial substance] encompasses all the worlds," he is reputed to have stated. Reasonably, Anaximander argued that this foreign matter changes into various substances familiar to us and that these in turn change into each other. Stressing the notion of change during the course of time, he declared, "Into that from which things take their rise they pass away once more, for they make reparation and

50

satisfaction for one another for their injustice according to the ordering of time." According to Bertrand Russell, this reference to justice refers to Anaximander's view that fire, earth, and water are constantly transforming, the proportions of each substance kept balanced by some undefined natural law.

Transformation and change had thus become concepts embedded in natural philosophy even some twenty-five centuries ago. Indeed, eternal motion—and even a primitive kind of evolutionary generation—were central to Anaximander's philosophy. Conceiving that Earth was not created but metamorphosed from the (unnamed) primal matter to an early fluid state, he claimed that all living creatures emerged from the moist land as the Sun gradually dried our planet. Remarkable for the time, Anaximander further taught that all animals descended from the fishes; as for humans, "aquatic men" first maneuvered like fish in the oceans and only later negotiated the land after some added progression.

Of course, by modern standards, Anaximander's somewhat anthropomorphic idea that humans emerged much as they are today is incompatible with what we now know, but his glimpse of progressive development from simpler to more complex structures contains at least a germ of our modern evolutionary principles. We can reason even further in this respect, surmising that Anaximander's fragmentary writings display a dim measure of survival or persistence against trying circumstances—an idea embraced primitively a century later by Empedocles and more formally a century ago by Darwin himself.

Interestingly enough, Anaximander did not confine his thoughts to life on Earth or to Earth itself. He hypothesized that it was eternal motion or change that gave rise to the origin of worlds—worlds plural, for he evidently regarded our planet as only one of many. He even speculated about the Sun, teaching that it was much larger than Earth—a novel yet unsupported statement at the time. But we remain unsure to what extent he believed that the Sun and stars also originated from his unfamiliar and unspecified primal matter.

One of Anaximander's pupils, Anaximenes (c. 588–524 B.C.), extended his mentor's ideas, specifically regarding life's origins and generally regarding the concepts of change and progression in nature. Anaximenes introduced the idea of primeval slime, a mix-

. . . living creatures emerged from the moist land . . .

ture of water and earth, from which, under the influence of the Sun's heat, plants, animals, and humans were spontaneously created. Although, as noted in the previous chapter, we now realize that such an abiogenic idea makes no sense scientifically, I regard Anaximenes' recipe as rather prophetic considering today's biochemical simulations of Earth's primordial soup.

More significant, Anaximenes reasoned that change inundates much of nature, not just the living world. Further, and of no small import, he suggested *how* things change. Claiming that air is of greater importance than either earth or water, Anaximenes taught that the condensation of air yields all substances; according to him, fire is rarefied air, but when condensed, air first becomes cloud, then water, earth, and finally stone—a clear progression from gas to liquid to solid. He even preached that the soul itself is made of air and thus postulated a capacious force present in everything everywhere. Metaphysically anticipating today's much-debated Gaia hypothesis—the idea that Earth's atmosphere is a biologically modulated circulatory system produced by the biosphere (a speculation to which I shall return in the Epilogue)—Anaximenes noted, "Just as our soul, being air, binds us together, so do breath and air encompass the whole world."

I judge the essence of Anaximenes' ideas to have considerable merit, for he taught that *all* things depend on the *changed* state of a primal substance (in this case, air). Of critical import, Anaximenes not only grasped Anaximander's progressionary (though not truly evolutionary) ideas but also extended them to include inanimate objects.

Among many other adherents of the Ionian school, Xenophanes (c. 576–480 B.C.) prominently shared some of these same ideas, speculating that human origins extend back to the time in Earth's history when the oceanic waters were drying to expose more land. Though extant fragments fail to specify how far he elaborated these ideas, we do know that Xenophanes preached a kind of spontaneous generation of life, with the Sun warming the Earth to produce both plants and animals. He is most famous in the annals of science for having been the first to recognize fossils as the decayed remains of organisms that once lived.

In these ways Milesian thought contributed much to the history of ideas largely because it forced a clean break with the past; its

53

open-minded philosophy of approach enabled an abandonment of answers given by tradition or myth in favor of a new instrument of investigation—reason. Having no strong religion or established priestly class, and maintaining good trade and social relations with Babylonia and Egypt, the Milesians used their rich cultural interaction to help weaken the Greek mind to prejudice and superstition remaining from Homeric days. Free inquiry emerged and began to flourish. To be sure, the Milesians viewed the world rather primitively, even naïvely, by modern standards, and hardly any of their hypotheses have withstood the test of time, but they asked good and basic questions, at which point I am reminded of a personal weltanschauung told me many times by one of my teachers, Harvard's Nobel physicist Edward Purcell: "Knowledge advances at the rate at which we ask correct questions."

As the first, pre-Socratic phase of Greek inquiry continued, efforts became less philosophical, more naturalistic. Leading Greek naturalists, chiefly Heraclitus and Empedocles, became far bolder and more fruitful in their contentions. Among them we find a strong notion of metamorphosis, and through extension of the Milesian ideas of spontaneous generation (abiogenesis), these naturalists anticipated surprisingly well the essence of evolution as developmental change.

In my view, the greatest of the early Greek naturalists was Heraclitus (c. 535–475 B.C.). Impressed with ceaselessness everywhere, he regarded perpetual change as the universal law. All things, he said, are in a state of flux, transposing into new shapes. Reality is changing, unstable; even the stillest matter harbors unseen flux and movement. Besides his most famous (and possibly apocryphal) quip noted earlier in Chapter 1 about "change is all," other relevant fragments (cited by Plato) include "The Sun is new every day" and "You cannot step twice into the same river, for fresh waters are ever flowing upon you." Perhaps Plato described Heraclitus best: "He taught that nothing is being, everything is becoming."

To Heraclitus, air, earth, and water were not as crucial as fire; for him, fire was the fundamental entity probably because, having the least consistency of all materials, fire best symbolized his central doctrine of perpetual change. Like a constantly fluttering flame (or, as mentioned in the previous chapter, like new stars triggered in the violent aftermath of supernovae), all things are born via the death of something else. What's more, Heraclitus seemingly elevated the role of fire to a "law" of nature. Uncannily, though superficially, resembling our modern laws of physics that theoretically operate at the very creation of the Universe—even governing that primal "nothing" extant prior to the emergence of energy, matter, and life (consult Chapter 4)—the law of Heraclitus assumed a permanent nature: "The world, the same for all things, no one of gods or men had made; but it always was, and is and shall ever be an everlasting fire, with measures kindling and measures going out."

Lest we infer the existence of Heraclitus' fire as unchanging, thereby violating his basic doctrine, let us recognize that many modern scholars have argued that fire is more of a process than a substance per se. In our modern cosmologies, aspects of energy have some permanence. Unlike matter, but like fire, energy is not a tangible property as much as a physical process. To be sure, as we shall see in Chapter 3, it is this process—the ever-present flow of energy through material structures—that helps construct and maintain all things, including galaxies, stars, planets, and life forms. Only one property in Heraclitean Fire was changeless: the burning process itself. Likewise, we now surmise only one property is likely changeless in cosmic evolution: the process of energy flow itself.

Not that Heraclitus apparently had any idea *how* things change; he was not a bona fide evolutionist. Some scholars in fact regard him as more metaphysician than natural philosopher. Admittedly, the fragments of Heraclitus ("the Obscure") are laced with mystical overtones: "All things arise from the one, and the one from all things; but the many have less reality than the one, which is God." Yet his recognition of nature's ranging breadth and of changes everywhere, including allusions to cosmic history as a repetitive cycle, comprise his principal contributions to the history of evolutionary thought. A specialist he was not! Rightly, no one can deny that his stress on the concept of *perpetual* change has had a lasting influence.

Empedocles (c. 492–432 B.C.) advanced beyond any of his predecessors and is often considered the father of the concept of evolution. More than a metaphysician, he made the first recorded observations in embryology and was also responsible for enunciating the time-honored maxim that all things are composed of air, earth, fire, and water. In a certain sense, Empedocles was a bit of a diplomat, a kind of synthesizer in his own time, honoring the water of Thales, the earth of Anaximander, the air of Anaximenes, and the fire of Heraclitus. These four "elements" he mixed in various ways, binding them with a force of "love" and separating them with a force of "hate." (Not surprisingly, with ideas like these, the subject of chemistry, quite advanced among Orientals, suffered a severe blow in the West; it seems that interest in chemical matters did not recover until the late Dark Ages, and only then when charlatanistic alchemists tried to change cheap metal into precious gold.) And like Heraclitus, Empedocles argued that every compound substance is temporary; only his four elements and two forces are everlasting.

Empedocles preached that nature produced lower and higher life forms neither simultaneously nor without effort; plants came first, animals later and only after a long series of trials. According to Empedocles, all organisms arose by means of the chancy interplay of the two great forces of nature upon the four elements. Animals initially appeared not as complete individuals but as parts of organisms—heads without bodies, arms devoid of hands, hands minus fingers, and so on. Gradually, with love triumphing over hate, various body parts began to seek each other and unite. With the forces causing a confusing interaction of body parts, all sorts of extraordinary beings can be imagined. Explaining nature in this way, Empedocles sought to understand not only the many varied life forms of our contemporary world but also the bizarre and monstrous creatures of Greek mythology, including, for example, beasts with human heads and humans with the heads of beasts.

Empedocles was at his best when writing of evolution, for he was much more of an "evolutionist" than any of his forebears. He actually seems to have anticipated a crude version of Darwin's major explanatory principle of natural selection, a kind of "survival of the fittest," yet Empedocles' view of selection was rather instantaneous, remotely like today's popular biological theory of punctuated equilibrium as an alternative to Darwinian gradualism. First, from fires within the Earth arose shapeless masses made of earth and water—sort of slimy, haphazard concoctions without limbs, reproductive organs, or speech capabilities. Later came the division of the sexes and the familiar mode of reproduction. He did not envisage these trials of nature as occurring within an organism and its subsequent offspring, thus improving each generation over the course of time—much as we now know biological evolution to work. Rather, he postulated a series of direct births from nature. Most of these organisms were unfit to live and became extinct; others, after ceaseless trials, were considered perfect by nature and thus survived.

Significantly, Empedocles stressed that changes in the material world occur by chance or purposelessness, not by design or godly intention (an attitude also championed by his leading contemporary, the atomist Democritus). Still, aside from his claimed interplay of love and hate, Empedocles lacked any genuine understanding of *how* life forms change—namely, how adaptations serve, over generations of time, to successively modify lower into higher forms. His "theory of evolution" just described is, of course, preposterous by modern standards, but his ideas do demonstrate that the Greeks nonetheless made progress regarding the central features of what one day would become genuine Darwinism. In effect, Empedocles revised the previously held abiogenesis idea of life's origin, making it more naturalistic, thereby granting his theory a semblance of modern science.

Despite his spearheading the first of all scientific revolutions, and one that highlighted the role of chance, Empedocles did not deny supernaturalism. Subscribing to a belief in gods, he apparently took his work too seriously and went one step beyond: He proclaimed himself to be one such god. Further, since he realized the need to test hypotheses, he even devised an experiment to verify his contention. To prove that he was divine, he supposedly jumped into the mouth of a volcano, a rather rash act memorialized by an anonymous poet:

> Great Empedocles, that ardent soul,
> Leapt into Etna and was roasted whole.

The pre-Socratics advanced magnificently the early development of evolutionary thinking; our species' first critical appraisal of nature could have hardly produced otherwise. Issues such as environmental change, lengthy time scales, spontaneous generation, and (ontogenetic) changes in individual life forms were broached within a rational and self-consistent context. But the pre-Socratics did not discover biological evolution as we now know the phenomenon, nor should they even be considered evolutionists per se. Alas, these earliest of philosophers were, perhaps like their precursor theologians, obsessed with understanding the origin of things more than their subsequent change.

Just as, in retrospect, the ancient Greeks seemed poised to develop a truly impressive system of nature, so their philosophy of approach altered drastically, turning, as if in recoil to rationalism, toward metaphysics and "higher authorities." The eminent twentieth-century evolutionary theorist Ernst Mayr has reasoned intriguingly that this harmful (to biology) retrogression was likely caused, ironically, by a competing science. The mathematical abstractions of the Pythagoreans and especially the increasing Greek preoccupation with geometrical order joined with other emerging cultural influences (such as a belief in an underlying fixity of things propagandized by the essentialist Plato) and induced intellectuals to embrace theism once again. Nonetheless, this odd sort of theism incorporated some naturalism, for even the most distinguished trio of philosophers to come could not ignore the materialistic advances of the pre-Socratics. The result was the formal inauguration of teleology: the doctrine that nature is (vitalistically) governed by design and/or purpose and that all material things have a final cause or reason for being.

> For hym was levere have at his beddes heed
> Twenty bookes, clad in blak or reed,
> Of Aristotle and his philosophie,
> Than robes riche, or fithele, or gay sautrie,
> But al be that he was a philosophre,
> Yet hadde he but litel gold in cofre.
> —Geoffrey Chaucer

The second stage of Greek thought—teleological in outlook, indeed one that claimed direct influence of a Grand Designer or "unmoved mover"—was dominated by perhaps the most influential philosopher of antiquity and one of the most synthesizing thinkers of all time. Aristotle (384–322 B.C.) towered above all his predecessors and by the force of his own genius created the subject of natural history. His observations of marine plants and animals during his boyhood near the seashore, coupled with his thorough familiarity with then-extant Greek philosophy, led him to grasp the notion of continuity or gradation. In fact, Aristotle postulated a *single* unbroken sequence of life forms—a grand ladder-shaped hierarchy of world phenomena termed a "Great Chain of Being" by later writers—extending from inanimate objects through plants to animals, including humans, a concept that remained intact until the introduction, in the nineteenth century, of multiple branching (bush- or tree-shaped) chains.

Aristotle's writings show that he clearly distinguished between plants and animals, as he did between the organic or living world and the inorganic or lifeless world. But he also fostered the abiogenic myth that not merely smaller but also larger animals, such as frogs and snakes, are produced spontaneously from the moist matter of the Earth. And he seems to have been overly influenced by Plato's antievolutionary (essentialist) orthodoxy, arguing for an immutability of species throughout nature.

Aristotle's works comprise a mixture of genius and error. His failures in descriptive science chiefly occurred when he departed from his own principles of verification, relying instead upon the informed hearsay of his day. This is an important point, for where Aristotle was correct, he was so largely because of experimentation and especially observation. Unlike Socrates, Plato, and many of his other philosophical forebears who relied mainly upon deductive logic as the means to true knowledge, Aristotle utilized the process of induction. He was therefore among the first of recorded history to use, systematically, the scientific method: a twofold procedure of theory and experimentation that comprises the central doctrine of all natural scientists today. Perhaps no one has better promoted the scientific method than Aristotle in his *History of Animals*—to wit, "We must not accept a general principle from logic alone, but must prove its application to each fact; for it is in facts that we must seek general principles, and these must always accord with facts. Experience furnishes the particular facts from which induction is the pathway to general laws." In short, he held that errors arise not because our senses are false media but because we mistakenly interpret their testimony.

Even so, Aristotle's theories regarding the origin and development of life went far beyond what he could have attained by the legitimate application of his professed investigative method. For he put his facts together into a system of nature having the teachings of Socrates and Plato as its primary philosophical basis. The result was a grand naturalistic synthesis, but one having clear metaphysical, even teleological overtones.

Aristotle envisioned a complete gradation in nature, a progressive development of life paralleling an ongoing but distinct ripening of the soul. (In doing so, I might add, he introduced an obscure dichotomy between matter and mind on which many of today's religious institutions are partly based.) The lowest stage in his *scala natura* is the inorganic, which passes into the organic by direct metamorphosis; in this way, matter changes into life. Plants are animate compared to minerals, but inanimate compared to animals; plants, Aristotle reasoned, have powers of nourishment and reproduction, but no feeling or sensibility. Next on his list of life's progression are simple marine creatures, such as sponges and sea anemones. Then come the animals with sensibility, possessing a desire for food and other necessities of life, and including locomotion to fulfill these desires. Not surprisingly, I suppose, given his tendencies toward anthropocentric teleology, at the pinnacle of Aristotle's one long and *continuous* hierarchy reside humans, who alone have the ability to generalize and form abstractions.

Throughout his work Aristotle was guided by what he felt was not only order in the Universe but perfection as well. According to his voluminous writings, nature proceeds constantly by the aid of gradual transitions from the imperfect to the perfect, growing steadily in complexity. The heart of his theory of nature is a so-called internal perfecting principle, a tendency that moves, even drives organisms progressively forward toward perfect types. The Universe and everything in it advance toward something continually better than what went before. All life forms tend toward humans, for we, in Aristotle's view, are the crowning end, sole purpose, and final cause. Thus, his theory was thoroughly egocentric: "Plants evidently exist for the sake of animals, and animals for the sake of humans; thus nature, which does nothing in vain, has done all things for the sake of man."

Let there be no misunderstanding here. Contrary to some popular beliefs, Aristotle was not a biological evolutionist as we know the term today. He envisioned no temporalization of his organiza-

. . . scala natura . . .

tional hierarchy from mineral to human; rather, his chain of being was a static ordering of unchanging, created entities—a set of creatures in rigid positions of an ascending structure that represented not time but the eternal order of things. As such, his chain resembled an eternally fixed ladder of immutable plants and animals spanning from minerals on the bottom rung to rational humans at its apex. In his system, changes occurred (aside from the familiar phenomenon of growth of given life forms) by means of successively new origins *within a given species*. Aristotle thereby embraced the notion of a history of change; indeed, he became the first great natural historian. But he held that like begets like, interpreted fossils as mere aberrations in rock strata, and rejected the possibility that any species of plant or animal had ever become extinct. Thus, Aristotle foresaw no change from one species to another, hence, no genuine evolution of existing types.

Aristotle went beyond describing *what* his graduated scheme was; he also outlined *how* it worked. But here he went far astray, opting to forsake the naturalistic inclinations of his predecessors for higher authority. Rejecting Empedocles' (Darwinian-like) contention that organisms result from a chancy survival within changing environments, Aristotle's theory had a decidedly metaphysical basis. Stating that "Nature does nothing without an aim," Aristotle felt compelled to postulate a Grand Designer as the primary cause of all things. Nothing, he held, which is regular, perfect, and beautiful can be the result of accident. Even so, Aristotle was not the consummate theist; he denied that a Prime Mover interfered with the daily routine of organisms. In his mind, the struggle toward perfection is a natural process: "It is due to the resistance of matter to form that nature can only rise by degrees from lower to higher types." Thus, he envisioned a spaceless, changeless, perfect, and eternal God who had ordained the laws of nature, but not one who intervened once the process had begun. Accordingly, Aristotle was a confirmed teleologist. Had he accepted Empedocles' hypothesis of nature's chancy struggle to adapt and survive, he would have been a prophet of Darwin.

What about Aristotle's thinking in the larger realm? Did the popular Greek concept of change extend beyond life forms? To be sure, he stated, "Nature makes so gradual a transition from the inanimate to the animate kingdom that the boundary lines separating them are indistinct and doubtful." For example, in his brief but elegant treatise on meteorology, Aristotle is full of penetrating insight, postulating among other things a cyclic world where the Sun evaporates seawater, dries up rivers, and eventually transforms oceans into bare rock; in turn, the uplifted moisture gathers into clouds, gradually renewing the rivers, lakes, and seas. Everywhere on Earth change is rampant, often imperceptible but effective.

Peculiarly, though, for I find it illogical given the above, Aristotle rejected change regarding the Universe generally. In his work *On the Heavens*, he set forth his straightforward notion that Earth is eternally fixed at the center of all things. Inside the orbit of the Moon, he said, objects are subject to generation and decay, but from the Moon outward, everything is fixed, ungenerated, and indestructible. Thus, the stars and planets, according to Aristotle, are unchanging, indeed unchangeable—a secure and harmonious worldview of pervasive order and contentment. Interestingly enough, while everything within the sublunar sphere was held to be made of air, earth, fire, and water, Aristotle claimed that the divine and perfectly spherical heavens beyond harbor a fifth (unspecified) element comprising the stars and planets; ironically, he failed to include fire in his theorized composition of the stars. All in all, he was quite wrong on virtually all celestial issues, which probably explains why I personally did not think highly of Aristotle the scientist until my recent research for this book increased my awareness of his manifest contributions to observational biology and earth science. Paradoxically, his myriad observations of nature laid the foundation for Darwin's grand (biological) evolutionary synthesis, but his personal antievolutionary stance decisively and wrongly influenced just about everyone for the next two thousand years. Not until Renaissance times (and in some cases not until our current century) were more objective and critical naturalists able to overthrow the Aristotelian worldview, thus establishing beyond reasonable doubt not only that Earth circles the central Sun but also that all material things change whether they be terrestrial or extraterrestrial.

> No single thing abides; but all things flow.
> Fragment to fragment clings—the things thus grow
> Until we know and name them. By degrees
> They melt, and are no more the things we know.
> —Lucretius

The third and final, or post-Aristotelian, phase of ancient Greek thought was less theistic, to a certain extent agnostic, or at least skeptical of an overall design. With the conquest of Greece (c. 330 B.C.) by the Macedonian Alexander the Great and the subsequent loss of national independence, speculation on the causes of nature became more restrained. The teachings of Epicurus (c. 341–270 B.C.) were representative of the time and of the widening gap between the theists and the agnostics. The chief intention of the Epicureans was to establish the principle of natural, not supernatural, phenomena. While some regard them as simply bent on destroying religion—Epicurus held the two greatest sources of fear to be death and religion—their materialistic attitude nonetheless served a useful purpose by protesting the increasing devotion of the times (and later among the Romans) to magic, astrology, and the occult.

At any rate, Epicurus and his associates contributed little to the idea of evolution, instead relying heavily on the previous arguments of, especially, Empedocles. At every point of his philosophy, Epicurus excluded teleology while embracing a mechanical view of nature. Nor was he even interested necessarily in truth; provided that any given phenomenon had one or more natural explanations (without resorting to the gods), he claimed it idle curiosity to try to decide between them. Opposing the increasing infiltration of Orientals preaching apathy and resignation in the face of defeat (championed initially by Zeno, the Stoic), Epicurus insisted that pleasure (as an antidote to defeat) guides the struggle for existence among the life forms of our world: "Nature leads every organism to prefer its own good to every other good. . . . We must not avoid pleasures, but we must select them."

Mechanistic thinking was in vogue among the intellectuals of the time, especially those who worked at or were familiar with the legendary library in the Hellenistic city of Alexandria, Egypt. The geometer Euclid (fl. c. 300 B.C.) produced his *Elements*, one of the most famous and uniformly correct books of antiquity, indeed, perhaps the crowning testimonial to the Greek intellect. The mathematician Apollonius (c. 261–? B.C.) derived the principles of conic sections. Eratosthenes (c. 276–194 B.C.), director of the Alexandrian library, used Euclid's geometry to measure the radius of the Earth; without precise instruments he came to within one percent of the correct value, finding the answer to be 6,350 kilometers, compared to today's average (spacecraft-measured) value of 6,371 kilometers. Archimedes (c. 287–212 B.C.), a Sicilian who had studied at Alexandria, undertook many of his famous hydrostatic experiments and postulated several original mathematical axioms. And somewhat later, among many others associated with the world's first true research institute, Hipparchus (fl. 146–127 B.C.), the foremost astronomer of antiquity, cataloged the positions and brightnesses of hundreds of stars, correctly anticipating that they come into being, slowly move during the course of centuries, and eventually perish.

Of much import generally, though not specifically regarding the history of the concept of change, Aristarchus of Samos (c. 310–230 B.C.) postulated that all the planets, including Earth, revolve around the Sun and, furthermore, that the Earth rotates on its axis once each twenty-four hours—nothing less than a complete heliocentric model nearly two thousand years before Copernicus, who usually gets most of the credit. (Actually, Aristarchus may have only elaborated on an earlier proposal made by Pythagoras.) In his conception of Earth as merely one among many planets orbiting through space, Aristarchus showed for that period an extraordinary emancipation from the anthropocentric consensus.

For the most part, though, the Greeks in the few centuries before Christ were not generalists or universal philosophers like their predecessors of earlier ages. Instead, they were specialists in the modern sense, content to be mathematicians, astronomers, and the like, yet apparently not aspiring to originate grand new worldviews.

Some two centuries thence, the Roman philosophical poet Titus Lucretius Carus (c. 99–55 B.C., a contemporary of Julius Caesar's) promoted the Epicurean mechanistic ideas, including the exclusion of theism. In what is perhaps the greatest scientific epic ever composed, On the Nature of Things refreshed the doctrines of Epicurus and, to a certain extent, those of Empedocles. Lucretius cogently sketched the origin and history of life on Earth, including the biological and cultural changes among humankind from prehistoric times through early hunter-gatherer and agricultural societies to Greco-Roman civilization. And he poetically proclaimed that all adaptations found in the Universe, both organic and inorganic, are merely special cases of the infinite possibilities of chancy mechanical events. Nothing exists, Lucretius lyricized, beyond atoms, space, and law; and the law of laws is evolution and dissipation everywhere. Consider some of his cosmic evolutionary insight:

> Globed from the atoms falling slow or swift
> I see the suns, I see the systems lift
> Their forms; and even the systems and the suns
> Shall go back slowly to the eternal drift.

> Thou too, oh earth—thine empires, lands and seas—
> Least, with thy stars, of all the galaxies,
> Globed from the drift like these, like these thou too
> Shalt go. Thou art going, hour by hour, like these.

> Nothing abides. Thy seas in delicate haze
> Go off; those mooned sands forsake their place;
> And where they are, shall other seas in turn
> Mow with their scythes of whiteness other bays.*

Such materialistic ideas well suited the early Roman temperament.

Later, under Roman domination during the first centuries after Christ, free inquiry among the Greeks was again slowed, and open speculation about the causes of things subsided. Despite the mechanistic, nonteleological bent of the Epicureans, Aristotle's influence proved too widespread, his disciples too numerous, his writings too comprehensive. The idea of the fixity of cosmic things, including Earth at the center of the Universe, regained popularity. And the theistic posture of Aristotle led to his being adopted by, and greatly influencing, the fathers of the early Christian Church. Ironically, the view that heavenly bodies are immutable, so cemented in the minds of medieval Christians, dates back well beyond the Romans and even past Aristotle to the pagan worship of the Sun, Moon, and planets as unchanging gods.

> I believe in one God,
> Creator of Heaven and Earth. . . .
> —from the Nicene Creed

Whatever the reason—perhaps hardly more than the Lucretian warning of agricultural decay caused by exhaustion of the soil—

*W. Mallock, tr., *Lucretius on Life and Death* (London: Black Publishers, 1900).

imperial Rome eventually became impoverished, its organization disintegrated. Order gave way to chaos everywhere as commerce declined, roads deteriorated, and cities capitulated, decade after decade, to the Germanic and Persian tribes crossing an unstable and ever-shrinking frontier. Though this is hardly the place to debate how or why, the grandeur that was the empire passed into the papacy, thus heralding a millennium of scripturally oriented philosophy.

As regards the principal objective of the two great phases of Western thought, Christianity differs profoundly from Greek philosophy. The Greek problem essentially concerned change (or "motion," the broad Greek-language translation of which includes not merely "movement" but also "change"), yet most Greeks failed to question the very existence of things, to challenge the establishment of being. Existence, however, even of matter itself, is precisely what Christians find puzzling; hence, their foremost concern with the origin of being or creation itself—a concern that could hardly be more clearly stated than in the first sentence of the Book of Genesis: *In principio creavit Deus caelum et terram.*

The idea of creation advanced in the Judeo-Christian age added a rather new concept to the progression of human thought. Arising largely from Old Testament Scripture, the creation ethos forces us to face squarely the institution of religion, as distinct from philosophy. While some of today's critics assert that philosophy and religion are not only incomparable but incommensurable, recall that my aim here is to explore the historical role played by the concept of change regardless of from whatever quarter it may have arisen. And to some scholars, creation manifests the ultimate of all possible changes. From any standpoint, religion has decidedly affected post-Aristotelian philosophy since much subsequent Western thought and European history derive from the Genesis passage noted above. Furthermore, as I shall explicitly treat in Chapter 4, most of modern science accords with a liberal interpretation of it.

In the West, Aurelius Augustine (A.D. 354–430) heads a long line of Christian scholars whose views were influenced by the need to agree with Holy Scripture. Still, this perhaps greatest of all Christian intellectuals (said by some to be the last ancient man and the first modern man) managed to impart a surprisingly naturalistic bent to sacred writ. For example, though Genesis proclaims an instantaneous creation of plants and animals, in Book XI of his incomparable autobiographical *Confessions,* Augustine interpreted this allegorically to mean a gradual development, much akin to Aristotle's progressionary model of imperfect to perfect. And again, after considering the question of time, Augustine further implored us not to regard the six days of the Genesis creation as equivalent to precisely six of our solar days. According to him, the Bible should be diagnosed as less of a literal and more of a symbolic representation of actual events or, as I have often personally maintained, as a poetic expression of potentially basic truths. Consequently, Augustine's fifth-century hermeneutics conform rather closely to the progressive ideas of today's pluralistically oriented theologians who accept the concept of evolution.

In Augustine's view, a single God—the one true being—created the Universe from nothing (*creatio ex nihilo,* not even from His own being) and endowed the cosmos with its basic properties and laws. Time itself, he said, began with the birth of the world; thus we apparently need not worry about what preceded the origin of the Universe. Not surprisingly, like any sage, Augustine failed to elaborate how something can be created from nothingness—a void— though as bizarre as it may seem, I shall clarify in Chapter 4 how modern physics appears poised to do just that.

I stress again that unlike the Greeks, who reasoned that matter itself is eternal and uncreatable and (with some exceptions) that a god or gods architecturally form and shape matter, Augustine followed the revelation of the Testament, maintaining that God created substance itself, not only order and arrangement. Further, he argued that much of God's creation involved granting nature the potential to produce organisms; hence the origin of Augustine's liberal scriptural interpretation that various objects and geometrical patterns surrounding us in the Universe originated gradually from formless, chaotic matter. All development thus follows a natural course of material changes dictated by God, the Creator. Even men and women are products of this natural development, though we have emerged "in the image of God" by grand design. Clearly, such views are heavily teleological and in no small way (though indirectly) influenced by the teachings of Aristotle. Even so, the rise to dominance of Christianity in the West meant that Aristotelianism was in eclipse.

For roughly a thousand years thereafter, original comment on the interpretation of Genesis nearly ceased and the hitherto steadily improving appreciation of the role of change in nature ground to a virtual halt. These were medieval times, including the secularly stagnant Dark Ages; a whole millennium—from the destruction of the Alexandrian library and the fall of the Roman Empire (c. A.D. 450) to the onset of the Italian Renaissance (c. 1450)—saw little advancement regarding the idea of evolution, or much other factual knowledge for that matter. Barbarian invasion was surely a factor, yet so was papal edict. The Church of the Middle Ages, at least until the twelfth century, was for the most part overtly hostile toward materialistic learning. ". . . the praises of Christ cannot find room in one mouth with the praises of Jupiter . . ." were the words of the "enlightened" Pope Gregory the Great in the sixth century.

Of course, men in cloisters and other centers of Western medieval culture debated medicine, ethics, probably politics, and especially religion, while largely seeking to perfect Augustinian thought. And they copied and compiled previous knowledge while maintaining an intellectual status quo. But scientific observation, even natural speculation were at a virtual standstill with little or no progress achieved.

Fortunately in the East mathematics and philosophy flourished, though the Arabs were more transmitters of established ideas than originators of new ones. Nonetheless, Islamic efforts served to preserve Greek thought, especially Aristotelianism, so when the West escaped the darkness, Greek views were relatively intact. The Mohammedans had managed to safeguard the essence of pragmatic civilization, leading to the rise of Scholasticism in the twelfth century and eventually the Renaissance some three centuries thereafter.

With the West's emergence from the Dark Ages around A.D. 1000 and the simultaneous rise of scholastic reasoning (at least where dogma had not made speculation too dangerous), Aristotle's writings became increasingly accepted as the supreme *secular* authority. As an apparently popular pastime, clerics and monks of that age devoted themselves to ever-refining reconciliations of Aristotle's teachings with those of Christ. In the process the intellectuals showed a distinct aversion to facts and rational inquiry, preferring instead protracted and pedantic controversies emphasizing verbal distinctions and semantic subtleties. How many angels *can* be balanced on the tip of a needle? And do these angels possess free will—a query to which I for one would rapidly respond in the negative, for who among us mortals or spirits would freely elect to perch in such a piercing locality? Interesting and entertaining intellectual polemics at best, I'm afraid.

In the main, pre-Renaissance Western thought culminated with the voluminous and systematic exposition of the greatest of all scholastic philosophers of the Roman Church, Thomas Aquinas (c. 1225–1274). An advocate of Aristotle, Aquinas, in his opus *Summa Theologiae*, leaned heavily on Augustine: ". . . some say that on the third day plants were actually produced, each in its kind—a view favored by the superficial reading of Scripture. But Augustine says that the Earth is then said to have brought forth grass and trees *causaliter;* that is, it then received power to produce them." This is typical of the Thomist writings that strove to establish truth by "legalistically" adapting Aristotle's Greek philosophy to standard Augustinian theology. The result was not a true synthesis, for as I have stated, Aquinas adapted Aristotelianism without submitting to it, the product being a masterful and precise set of teachings that were as influential in Christian circles then as they are dogmatic in Catholic institutions now. As regards the idea of evolution per se, Aquinas seemingly contributed nothing, embracing instead the Aristotelian fallacy of the immutability of species. And in one of his famous "proofs of God," Aquinas strictly rejected any element of chance.

> But the world is slipping away; the polished sky
> Gives nothing to grip on; clicked from the knuckle
> The marble rolls along the gutter of time—
> Earth, star and galaxy
> Shifting their place in space.
> —Norman Nicholson

With the onset of the fifteenth-century Renaissance, educated people began substituting the authority of the ancients for that of the Church, thus prompting a significant break from the static Christian worldview grounded in the concepts of Aristotle, Augustine, and Aquinas. But since the ancients disagreed with each

70

other, reasoned judgment was required to decide which of them to follow. This alone was a step toward emancipation—a need to think critically. In contrast with the cloistered meditation of the Middle Ages, intellectual activity became a delightful social adventure. University men were liberated from the narrowness of medieval culture; secular learning was on the rise. In particular, the scientific method—ideally a three-step process of observing or speculating, theorizing, and testing—became standardized, with experiments having become an essential part of the process of inquiry.

A renaissance it was because almost instantaneously, at least in Italy, where it began, Western scholarship converted from the passivity of theological dogmatism to the activity of rational inquiry and empiricism. Among many others, Leonardo da Vinci (1452–1519) stands out as perhaps the broadest thinker and tinkerer of the time, and not just in art and anatomy, for which he is best remembered. A genuine natural philosopher, da Vinci conducted in the Alps extensive paleontological studies of marine fossils and geological scrutiny of the rock strata in which they are formed. His observational evidence was compelling: The surface of Earth had not been eternally unchanged throughout history; rather, as da Vinci reasoned, time is the evil destroyer of all things (an intriguing notion akin to our modern ideas regarding entropy; cf. Chapter 3). At one and the same time his methods of inquiry were a far cry from those of the medieval period, and his conclusions equally far removed from the secure and content fixity that is the hallmark of the Aristotelian model of antiquity. Clearly da Vinci took change and history seriously, like Heraclitus well before him, maintaining that all reality is in a state of continuous flux.

Leonardo's penetrating insight was not confined to terrestrial issues. In his dynamic cosmology, he rejected an Earth-centered model of the Universe, preferring early in his life a mechanical interpretation of the cosmos. This was hardly a surprising position, in view of his penchant for designs of aircraft, submarines, armored tanks, and giant telescopes, among many other technical gadgets not actually realized until the present century. Later, in his mature years, he softened his mechanistic attitude, adopting instead an organismic model of the cosmos: Da Vinci held that Earth itself is a living being, a fascinating proposition to which I return in the Epilogue.

71

Another premier contributor to the early Renaissance was instrumental in decentralizing planet Earth and endowing it with a cosmic mundaneness that even today is symbolized by the term "Copernican principle." While he was modeling observations of the planets, the search for simplicity guided the Pole Nicholas Copernicus (1473–1543) to hypothesize the Sun's central position in the Universe. His heliocentric model, as enunciated in *On the Revolutions of Celestial Orbits,* published in the year of his death, had little influence in his own time, as the (then somewhat liberal) Church regarded it as only one of many theories. Ironically, the Copernican model became more influential nearly a century later, when, at the time of Galileo's condemnation (1633), Copernicus' works were judged heretical by the Catholic Inquisition. And as often goes unstated, this same revolutionary idea of heliocentricity was damned even more vehemently by leaders of the Protestant Reformation. Not only was a noncentral Earth contrary to a literal interpretation of Holy Scripture, but such a heliocentric system required the Earth to *change* its position—in short, to move through space. The stability and immutability of Earth were lost forever, and most people of that age didn't like it. More to the point of this chapter, the coming of Protestantism was a definite setback for evolutionism because it led to a total rejection of Augustine in favor of a reinforced authority of the Bible. The result was a rigid, literal interpretation of the "Word"—namely, fundamentalism. Here, then, is the source of modern science's difficulties with today's fundamentalists.

Even Copernicus was biased by the Greeks of old, basing his heliocentric model on the circle, the perfect geometrical figure of antiquity. Surprisingly, he was aware enough of early Greek geometry to be swayed by it yet evidently unaware of Aristarchus' earlier heliocentric model of our planetary system. Actually I have a strong (though unsubstantiated) suspicion that Copernicus, like Darwin, among many other acclaimed revolutionary scientists, may well have been influenced by early Greek ideas a good deal more than he admitted in writing. At any rate, of no minor significance, the advance made popular by Copernicus removed Earth from its universal preeminence, thus making less tenable the cosmic pinnacle assigned to us humans by most theological dogma.

72

The Copernican model was further simplified when the German Johannes Kepler (1571–1630) empirically introduced the notion of *elliptical* planetary orbits, a mathematical proposal made easier by the experimental proof then recently provided by Italy's Galileo Galilei (1564–1642) that projectiles follow parabolic paths. Less than a century thereafter the Englishman Isaac Newton (1642–1727) used his newly developed theory of gravitation to justify mathematically the movement not only of Earth but also of the other planets and moons of our Solar System. The substitution of ellipses for circles was no small advance; it amounted to an abandonment of an aesthetic bias that had governed astronomy since early Greek antiquity. As in any age, deeply ingrained prejudices were not easily discarded. In fact, the concept of heliocentricity was not proved to virtually everyone's satisfaction until a century and a half ago, when the German astronomer-mathematician Friedrich Bessel (1784–1846) first succeeded in measuring stellar parallax—the apparent oscillatory movement of nearby stars artificially caused by Earth's annual trek around the Sun. Thus, nearly three centuries were required to confirm objectively the Copernican idea of heliocentricity, but that objectivity *did in fact* eventually emerge to reveal reality.

The Copernican episode is a good example of how the scientific method, though affected at any given time by the subjective whims and human values of various researchers, does lead to a definite degree of objectivity—this despite shrill appeals to the contrary by sociologists and other practitioners of social "science" (which is in fact, and likely always will be, value-laden). Over the course of time many groups of researchers checking, confirming, and refining experimental tests will neutralize the subjectivism of individual workers. Usually one generation of researchers can bring much objectivity to bear on a problem, though some particularly revolutionary concepts can become swamped for long periods by cultural and institutional biases such as tradition, dogmatism, and even politics. Heliocentricity received objective confirmation nearly three centuries after Copernicus had popularized it and a couple of millennia after Aristarchus had proposed it. Numerous other such episodes abound. Galileo's physics and astronomy were overwhelmed when the counterreforming Church tried to preserve the status quo, and Lysenko's Lamarckian genetics was promoted until relatively recently by Soviet ideology for vested sociopolitical interests. Eventually, however, objectivism prevailed.

Permit me to make here an additional clarification regarding the scientific method. No natural scientist of my acquaintance claims our modern method reveals "truth." We do not even seek final or ultimate explanations, for most of us realize that truth is likely to be fundamentally unattainable if only because we would have no means of recognizing it as such. As the pre-Socratic, Xenophanes of Colophon, articulated perceptively, ". . . as for certain truth, no man has seen it, nor will there ever be a man who knows about the gods and about all the things I mention. For if he succeeds to the full in saying what is completely true, he himself is unaware of it. . . ." Rather, we today maintain, with an open mind and a readiness to change our theories during the course of numerous experimental and observational tests, that nature yields a certain measure of objectivity through unraveled facts, thus granting a progressively better "approximation of reality." It is in this sense that we claim to have made progress, both in quantitative terms of a greater amount of knowledge and in qualitative terms of a fuller, more accurate knowledge.

Now, an event of some cosmic-evolutionary significance occurred in the late Renaissance. In 1572 a new object glowed brightly in the heavens where the sky charts previously showed no such beacon. We now recognize this new object to have been a supernova, the brilliant aftermath of a massive star that ended its stellar evolutionary cycle by exploding. Another star similarly detonated in our Milky Way and was seen in the year 1604. (We can still observe and study their diffuse remnants some four hundred years later.) Together these events served to demolish the (then still popular) Aristotelian premise that everything beyond the Moon is immutable. The cosmos does change, albeit often slowly, and direct observational evidence was then at hand to prove it. Comets, too, provided Renaissance men and women with ample evidence that celestial change is widespread. (Before Renaissance times Aristotelians were forced to conclude incorrectly that the obviously changing comets were manifestations of sublunar phenomena.)

Similarly, the power of the medieval Church to ignore what was obvious in favor of its Aristotelian insistence on the fixity of cosmic things may be exemplified by the case of an earlier super-nova. As noted in the previous chapter, relatively recent studies have historically documented the sighting of a blazing supernova in A.D. 1054. Known as the Crab Nebula, since its appearance somewhat resembles that type of marine creature, the resulting (and still observed) debris was reported at the time by Chinese observers to have initially rivaled the brightness of the full Moon and to have been visible in broad daylight for many weeks. Middle Eastern Arabian astronomers also sighted this phenomenon and recorded their findings, as did the American Indians who left engravings of the event in the rocks of southwestern United States. By contrast, I have never been able to find any evidence that medieval European observers took note of the supernova, yet there is no possibility that its brilliance could have evaded European skies. Was the power of faith then so supreme that the populace uniformly ignored this great light show, or did the Church scribes excise all trace of the cosmic mutant from the historical record lest Aristotle be tainted? A most telling episode showing apparently how religion can literally blind science; "Such evil deeds could religion prompt" was the way Lucretius had put it a millennium earlier.

Aristotle's views on the immutability of deep space had prevailed for some two thousand years. Logic alone had wrongly dominated observation. Consequently, though few contemporary philosophers like to hear it said, Aristotle's ideas on the Universe had proved an enormous obstacle to progress. Astronomical evidence for sudden changes (supernovae), irregular motions (retrograde planetary paths), and physical imperfections (sunspots) were to lead the way toward a serious study of not only change in the heavens but also transformation both *of* Earth and *on* Earth.

Moreover, the idea of purpose, which since Aristotle's time had formed an integral part of the nature of our world, was removed from the fabric of what would become known as "science." The heavens might truly exist to declare the greater glory of God, but surely this belief could not interfere with a mathematical calculation or an astronomical observation. With the rapid rise and spectacular success of reasonably objective Renaissance investigation, we might suppose teleology to have played no further role in scientific explication. Not so.

The late sixteenth century also saw the emergence of perhaps the most rationalistic thinker to (that) date, at least among theologians. The Italian Giordano Bruno (1548–1600) combined Greek influence, Arabian philosophy, and even a bit of Oriental mysticism to fashion a speculative and rather ambiguous system of natural philosophy. He taught that a Universal Force conceived a graduated scale of organized beings, with each form the starting point for the next—a seemingly temporalized version of the Aristotelian hierarchy. And like Aristotle, he placed greater emphasis on reason than on observation. Consider Bruno's chief maxim (though here he may have been rebuking orthodox religion as much as the need to test): "The investigation of nature in the unbiased light of reason is our only guide to truth." So, even as recently as four hundred years ago, the leading proponent of the concept of change was a scientific rationalist, all right, but apparently not one who subscribed to our modern emphasis on experimental testing.

Admittedly, again as did Aristotle, Bruno claimed that eternal change is not purposeless but always occurs so as to eliminate defects. However, to what extent he might have been an evolutionist in the modern sense, we cannot be sure. With his notion of endless change as an agent of perfection within organic matter, Bruno seems to have been a follower of Empedocles and thus another forerunner of Darwin. And he may have at least glimpsed the essence of organic evolution and the historical transformation of all life on Earth, for his writings imply a development of simple life forms into complex organisms. However, the many Aristotelianisms permeating his works also imply that, though evolution relies on endless changes in matter, these changes are outward expressions of an overall design (hence the ambiguity noted above regarding his real stance on the rationalism of these issues, yet he may have been simply dodging the ecclesiastical authorities of the time).

Though spectacularly enlightening the arts and sciences, Renaissance advances were troubling for the Church. Amidst the Protestant revolt and the influx of Islamic religion into Europe (especially Spain), the Counterreformation strove to quash all liberal tendencies and produced some of the most conservative theologians of any age. Typical of the period, and a contemporary of Bruno, was Francisco Suárez (1548–1617), a Spanish Jesuit, who held a most literal interpretation of the Scriptures. Many consider him the modern father of Special Creationism, a view of spontaneous generation favored even today by many biblical fundamentalists in the United States. Rejecting the progressive attitudes of Augustine and Aquinas, Suárez argued that a "day" in the Testament refers precisely to our solar day of twenty-four hours and, further, that all things were created intact in the interval of six actual Earth days.

Despite the authority and esteem granted even today toward the teachings of Augustine and Aquinas, virtually all denominations of Christian theologians from the mid-sixteenth to the mid-nineteenth century departed from the original philosophical standards of the early Roman Church. With Aristotelianism on the decline, the rulers of the fragmented Church rejected Augustine's broad and allegorical interpretations of creation, embracing instead the Special Creation ideas championed by Suárez and other archconservatives. The Roman Inquisition solidified Creationism as a hallmark of the Church's teachings and, despite his mildly teleological views, ordered Bruno burned at the stake in 1600.

Augustine's somewhat liberal theology, particularly those parts permitting poetic interpretation of Scripture, had been eclipsed. Ecclesiastical authority was again challenging rational inquiry. Ironically, had Augustinian orthodoxy remained the official teaching of the Church, the idea of evolution might have been established far earlier than it was, and I daresay we would not likely be burdened with today's two-culture schism between religion and science. At any rate, until Darwin's pronouncement of his decidedly non-teleological views of nature raised the subject for all nineteenth-century society, it fell by default to the natural philosophers to be the guardians of the concept of evolution.

> All members develop themselves according to eternal laws,
> And the rarest form mysteriously preserves the primitive type.
> Form, therefore, determines the animal's way of life,
> And in turn the way of life powerfully reacts upon all form.
> Thus the orderly growth of form is seen to hold
> Whilst yielding to change from externally acting causes.
> —Johann Wolfgang von Goethe

The natural philosophers of post-Renaissance times contributed mightily to the nature of rational inquiry. Of foremost value, they championed the experimental test as a necessary supplement to thinking, thus fully grasping the inductive scientific method used by all natural scientists today. And they stressed the principle of natural causation, seeking materialistic causes for the development of our world, not supernatural ones that presuppose the incessant intervention of a God or gods. Such a Supreme Being was still theorized as the originator of all that exists, but subsequent to creation all events were considered regulated by natural "secondary causes" best exemplified by the mechanistic rules of the physical sciences—a deistic, though neither theistic nor atheistic, stance. In fact, so materialistic had the new post-Renaissance paradigm become that the Universe was taken to be a machine kept running by a set of general laws grounded in mathematics. The paramount goal of science, though it was still not yet called that, aimed to discover those laws and therefore bring all natural phenomena within our quantitative understanding.

The Englishman Francis Bacon (1561–1626), often dubbed the father of the scientific method, perhaps best elucidated the modern inductive method of investigation, whereby we reason from the particulars to the whole. Bacon roundly criticized the Greeks for their deductive methods, though I think somewhat unfairly since even Aristotle was an occasional user of the scientific method. (Noting, during a lunar eclipse in the fourth century before Christ, the curved shape of Earth's shadow cast onto the Moon's surface, Aristotle theorized that planet Earth is round. Further, he devised a test of his idea—namely, that at different orientations different parts of Earth's shadow projected onto the Moon should always be curved. Finally, he cited subsequent lunar eclipses as confirmation of his hypothesis that the Earth is indeed spherical.)

Bacon sought to discredit any approach based solely on logic. He stressed the scientific inductive method that builds models based largely on meticulous and critical observations, a method that, as I have noted earlier in this chapter, is designed to achieve over the course of generations some degree of objectivity by effectively removing the observer from the observed. We need not dwell too long here on the power of the scientific method; suffice to say that its widespread acceptance and use beginning early in the seventeenth century are the greatest methodological contribution to science made by the European Renaissance.

". . . in Flying Fishes, between fishes and birds . . ."

Bacon clearly recognized change throughout nature. Further, his prudent observations demonstrated *variations* in that change within a given species of life forms. We now realize that the accumulation of such variations can cause the mutability of species, a somewhat subtle feature central to modern biological evolution but one that was apparently overlooked or at least not well documented in antiquity. Thereby anticipating modern genetics (if not genetic engineering), Bacon was quick to suggest that humans should be able to experimentally induce variations among nature's myriad life forms. What is more, in Book II of his classic *New Organism*, he noted nature's many transitional forms spanning different types of species: "We have examples of them in Moss, which is something between putrescence and plants; . . . in Flying Fishes, between fishes and birds; and in Bats, between birds and quadrupeds."

Significantly, regarding the grander scenario of cosmic evolution—my avowed focus in the present work—Bacon extended his observations of change to include inanimate objects, especially ones distant from our planet. For example, in his principal tome, just cited, he illustrated a further instance of transitional form "in some comets, which hold a place between stars and ignited meteors." Throughout much of his writings Bacon espoused the big picture, an attitude captured well in one of his principal aphorisms: "The Universe is not to be narrowed down to the limits of the understanding, which has been man's practice up to now, but the understanding must be stretched and enlarged to take in the image of the Universe as it is discovered."

To this point in history, I rank Bacon second only to Heraclitus (to be later surpassed in my estimation by Kant, Chambers, Darwin, and Shapley) as having contributed originally and expansively to the foundations of modern *cosmic* evolution. Not that Bacon was a genuine evolutionist by today's standards, but he extended the Heraclitean doctrine of "eternal change" by noting many intricacies among those changes and by speculating, however tentatively, about the ever-changing nature of objects beyond our Earth.

In France René Descartes (1596–1650), deemed by many to be the founder of modern philosophy, helped further widen the gap

80

between those preferring to believe in literal interpretations of Scripture and those demanding observational and experimental tests of rational ideas. Philosophically, though not scientifically, Descartes boldly aspired to explain *all* things according to some (undefined) natural principle. Truly a man of my heart, he strove for a grand synthesis, the likes of which hadn't been attempted since Aristotle's proposed hierarchy of nature's multitudinous objects. Perhaps the ultimate determinist, Descartes theorized, but of course could not prove, that the whole physical Universe, including life forms, is mechanistic and thus should be rigidly describable in terms of a small number of physical laws, divinely given. Such a claim that the Universe is basically a vast machine required no small amount of moral courage in a country where the prevailing view of spontaneous creation had become prescribed dogma. Descartes is at his greatest, though without detail, in his *Principles of Philosophy* (1644):

> All the same, if we can imagine a few intelligible and simple principles upon which the stars, and earth, and all the visible world might have been produced . . . we reach a better understanding of the nature of all things than if we describe simply how things now are, or how we believe them to have been, created.

And again Descartes, a mathematician by trade, made clear his sweeping determinism and wholehearted acceptance of a mechanical worldview:

> As I considered the matter carefully, it gradually came to light that all those matters only are referred to mathematics in which order and measurement are investigated, and that it makes no difference whether it be in numbers, figures, stars, sounds, or any other object that the question of measurement arises. . . . I saw, consequently, that there must be some general science to explain that element as a whole which gives rise to problems about order and measurement. This I perceived was called universal mathematics. Such a science should contain the rudiments of human reason, and its province ought to extend to the eliciting of true results in every subject.

As for evolutionism specifically, Descartes seemed to straddle the issue. While proclaiming in much of his writings that nothing created perfectly by an omnipotent God can possibly evolve, he at other times showed ambivalence by presuming that an indifferent God could have manifested His perfection by merely creating the laws of physics, after which the world developed and evolved naturally to the stage we now know it. Descartes even speculated about the rudiments of cosmic evolution, suggesting that the operation of natural laws caused the ordered stars to emerge from a state of primordial chaos that was "as disordered as the poets could ever imagine." The Earth itself he took to be a highly evolved dwarf star. He then almost immediately repudiated these ideas (possibly to guard against charges of heresy), adding that "we know perfectly well that they [the Earth and stars] never did arise in this way."

My amazement never ceases at the apparently ambiguous scholarship of many of history's "great" intellectuals. Come to think of it, lack of clarity among academics, especially in scientific exposition, is one of the foremost problems in scholarship today. On the other hand, perhaps I am being overly harsh since a person's ideas might well change over the course of a lifetime.

The deterministic ideal was bolstered somewhat by the Britisher Isaac Newton, who himself had nothing appreciable to offer regarding the concept of change in nature but who nonetheless discovered the specific mathematics needed to justify the mechanical world paradigm of Bacon and Descartes. In a synthesizing step of first magnitude, he argued that a single law could explain the fall of an apple from a tree as well as the motion of a planet around the Sun. He thus concluded that the force of gravity that governs objects on Earth rules the wider cosmos beyond. To Newton, there is order in the Universe, and as Descartes had earlier philosophized, that order could be ascertained by mathematical formulae backed by scientific observations. In effect, Newton supplied the inorganic world with a set of regulating principles.

(Incidentally, I am intrigued by the cyclical way that researchers have viewed throughout history the interaction among material objects in the Universe. The ancient Greeks regarded geometry so highly that they demanded, for example, that planets describe perfect circles in their various orbits. Newton discarded this emphasis on geometry by introducing the notion of the force field, postulating that the Sun and planets are bound by the mutual attraction mediated by the invisible force of gravity. In turn, Einstein's relativity theory declared that gravitational forces are essentially nonsensical, instead interpreting such material interactions within the context of curved spacetime, a subject known to workers in the field by the tongue-twisting name of geometrodynamics. And now, in the most current parlance of quantum field theory, which we shall encounter in Chapter 4, physicists have returned to the concept of invisible, microscopic fields to describe the forces ruling our multifarious cosmos.)

The Dutch rationalist philosopher Baruch Spinoza (1632–1677) further extended determinism by demanding a natural cause or absolute logical necessity for all things. Here he is in a relevant passage from his *Ethics:* "The natural laws and principles by which all things are made and some forms are changed into others are everywhere and through all time the same." However, unlike his contemporary Descartes who admitted three substances—God, mind, and matter—Spinoza maintained only one: "God or Nature." Accordingly he embraced an undiluted pantheism, whereby everywhere God and nature are synonymous; the better you understand the material Universe, the closer you come to God. For Spinoza, there existed neither free will in humans nor random chance in the physical world. This viewpoint was made famous in the present century when Einstein, also rejecting chance, declared, "I believe in Spinoza's God who reveals himself in the harmony of all that exists." Even so, almost all of today's physicists concede the validity of the subject of quantum mechanics (cf. Chapter 4), wherein the element of irrational chance does play some definite, though indeterminate, role in nature.

The earliest unambiguous suggestion of *modern* evolutionism may well have been posited by the first of a long line of distinguished German philosophers Gottfried Leibniz (1646–1716), who, incidentally, shares credit with Newton for the (independent) invention of the differential and integral calculus. In his *Protogaea* (1693), Leibniz borrowed from the teachings of Aristotle to postulate a continuous, linear chain of organic beings: "All natural orders of beings present but a single chain, in which the different classes of animals, like so many rings, are so closely united that it is impossible either by observations or imagination to determine where one ends or begins." Yet unlike Aristotle and apparently unbridled by biblical chronology, he temporalized this chain. In a telling passage rich on several counts, Leibniz outlined his progressive views regarding the myriad (perhaps nearly infinite) gradations among the different classes of life forms, an idea doubtlessly influenced by the work of Bacon, Descartes, and perhaps Bruno (as well as conceivably by the infinitesimally incremental nature of the calculus):

> Everything advances by degrees in nature, and nothing by leaps, and this law controlling changes is part of my doctrine of continuity. Although there may exist in some other world species intermediate between man and the apes, nature has thought it best to remove them from us, in order to establish our superiority beyond question. I speak of intermediate species, and by no means limit myself to those leading to man. I strongly approve of the research for analogies; plants, animals, and comparative anatomy will increase these analogies, especially when we are able to take advantage of the microscope more than at present.

Later in the same work Leibniz argued, albeit somewhat metaphysically, that numerous life forms had flourished in earlier geological periods and had since become extinct. He even advanced a necessary prerequisite for genuine evolutionary thought: speciation by means of environmental change. "Indeed it is credible that by means of such great changes (of habitat) even the species of animals have been many times transformed," he wrote in reference to the relationship of the fossilized ammonites to the living nautilus. The lingering Aristotelian concept of the immutability of species, in vogue for some two millennia, was in full retreat. More fundamentally, the essentialism of Plato was explicitly rejected. The intellectual milieu itself was changing.

Leibniz notwithstanding, Immanuel Kant (1724–1804) is often judged the greatest German natural philosopher; in my estimation, he was the foremost of all the natural philosophers of the Modern Enlightenment, an eighteenth-century period of Western intellectual prosperity reminiscent of the pre-Socratic age and the Italian Renaissance. Kant was instrumental in propagating the Heraclitean ideal of "change everywhere" in advance of the first evolutionary naturalists or "scientists" late in the eighteenth century. Though completely pre-Darwinian, his writings display not only the idea of change but also key evolutionary words, such as "selection," "adaptation," "environment," and "inheritance." What's more, Kant was the first to explain logically and consistently that the Universe as a whole has evolved.

Kant's early writings, notably *Universal Natural History and Theory of the Heavens* (1755), demonstrate his comprehensive view of evolution, contending that all observable phenomena lie within the domain of natural causes, and thus that the origin and development of all things are explicable through the use of chanceless laws of nature. He wrote of tracing back all the higher forms of life to simpler elementary forms. And he clearly placed humans among the ranks of nature, alluding to our former quadrupedal posture and outlining some environmentally induced changes that were produced in us over the course of time. Furthermore, Kant introduced the now popular idea that all things began with a chaotic universal nebula out of which galaxies, stars, and planets gradually emerged from swirling, contracting vortices over periods of time considerably longer than anyone had previously dared imagine. In fact, Kant thought in terms of infinity of time: "The future succession of time, by which eternity is unexhausted, will entirely animate the whole range of Space. . . . The Creation is never finished or complete. It did indeed once have a beginning, but it will never cease." Clearly his was no longer a static cosmos but a dynamic, continuously changing one—a truly revolutionary, even heretical declaration made hardly more than two hundred years ago.

In his later life Kant somewhat cryptically backed off, dividing nature into the inorganic world wherein natural, mechanistic causes prevail, and the organic world wherein some sort of hidden, teleological principle dominates. In his three-volume set *Critique of Pure Reason, Critique of Practical Reason,* and *Critique of Judgment* (1781–90), he drew a sharp distinction between matter and life, vehemently rejecting the emerging theory of biological evolution. Claiming that science would never understand the growth and development of even a blade of grass, Kant judged it overwhelmingly impossible that humans would ever attain a successful explanation of the natural laws governing the changes among life forms. Less than a century later Darwin provided just such an explanation.

Such strident warnings of the limits to our knowledge—in Kant's case, in order to substantiate the practical need for moral theology—I find peculiarly associated with elderly scholars. Even today professors emeriti seem to be a good deal more metaphysical (and even occasionally teleological) than younger colleagues or than they themselves were in their youth. Whether it's wisdom or senility, I have not yet been able to judge.

> Organic Life beneath the shoreless waves
> Was born and nurs'd in Ocean's pearly caves;
> First forms minute, unseen by spheric glass,
> Move on the mud, or pierce the watery mass;
> These, as successive generations bloom,
> New powers acquire and larger limbs assume;
> Whence countless groups of vegetation spring,
> And breathing realms of fin and feet and wing.
> —Erasmus Darwin

In consort with the aforementioned philosophers, several naturalists of the eighteenth and nineteenth centuries contributed much to the modern idea of evolution, including the concept of cosmic evolution broadly conceived. While the philosophers essentially safeguarded the notion of change from attacks of the Church in post-Renaissance times (for spontaneous creation still dominated public opinion throughout this period), the naturalists began examining nature with increasingly fine precision. Of course, mid-eighteenth-century workers had few of the technological tools now at our disposal, yet those naturalists we now regard as eminent clearly embraced the scientific method, including its emphasis on observational and experimental tests.

Prominent among the early pre-Darwinian naturalists was the Frenchman Georges Buffon (1707–1788), though in his day his genius was not appreciated. His (then) radical views, especially as to the mutability of species, were too novel—apparently yet another example of intellectual inertia blocking ideas too advanced for the time. Nor were his ideas capable of proof in facts then discovered or tests then possible. By broadly illustrating the relationship between changes of environment and the transience of species, Buffon laid the basis for modern evolutionary zoology and botany; he is generally credited with having founded the subject of biogeography. Furthermore, in his forty-four-volume literary opus, *Natural History: General and Particular* (1749–1804), his far-reaching generalizations advanced considerably beyond the ancient Greek and early post-Renaissance philosophers, raising to respectability the study of evolutionism as a science.

Here is Buffon proposing his idea of the frequent mutability of species triggered by environmental change:

> How many species, being perfected or degenerated by the great changes in land and sea, by the favors or disfavors of nature, by food, by the prolonged influences of climate, contrary or favorable, are no longer what they formerly were? . . . One is surprised at the rapidity with which species vary, and the facility with which they lose their primitive characteristics in assuming new forms.

A strict naturalist or not—we cannot be certain—Buffon seemed unable to abandon the possibility that each species originated at the hands of a Creator, after which the observed variations in life forms were the effect of mere departures from those original types. "The deeper I penetrate into the depths of nature, the more I admire and profoundly respect her author." Like others before him, he constantly wavered between a belief in Genesis and the evidence of nature. Accordingly, though Buffon teetered on the brink of evolutionism all his life, his share in the development of the modern idea of evolution seems destined to be forever clouded.

Still, I'm impressed with Buffon's discussions of the struggle for existence, the elimination of unperfected species, and the contest between the fertility of certain species and their constant destruction. To me, Buffon not only championed the concept of environmental influence on life's changes but also anticipated, at least in some small measure, Darwin's later Malthusian-inspired idea of life's struggle for existence and even perhaps his principle of natural selection. Unfortunately Buffon failed to offer a specific explanation for the transformation of species, so the magnitude and validity of his contributions are mired in one's subjective interpreta-

". . . being perfected or degenerated by the great changes in land and sea, by the favors or disfavors of nature . . ."

tion of his voluminous writings. There he often contradicts himself, probably, as with others, in a bid to placate ecclesiastical authorities (especially at the Sorbonne) who still held pockets of power in eighteenth-century Europe.

Admittedly, a crucial flaw in Buffon's evolutionary thinking was his claim that the environment acts *directly*—that is, rapidly— during the lifetime of a given individual to change the structure of plants and animals. As we now recognize, environmental events (cosmic rays, drugs, etc.) affect the molecules (DNA) within organisms, producing rather slow and indirect structural alterations, for these molecular changes are inherited over the course of many generations.

Such errors are often balanced by trenchant passages like this one about the common ancestry of species: "If it were admitted that the ass is of the family of the horse, and different from the horse only because it has varied from the original form, one could equally well say that the ape is of the family of man, that he is degenerate man, and that man and ape have a common origin." Buffon reasoned further that if we extend this idea back to its logical conclusion, then all animals, and plants as well, must comprise but a single family. Alas, he characteristically retreated later in the same paragraph, invoking the benefits of direct creation and thus left the reader confused once again about where he really stood.

Buffon's evolutionary scenario did not end with life on our planet. He speculated about life on other worlds and even estimated, on the basis of geological cooling rates, likely time scales for the origin of life on such distant planets. He further proposed that over the course of history planets change from one condition to another. In his *Epochs of Nature* (1780), he laid out a natural history of the Earth that uncannily resembles the broad outlines of our current ecological and biogeographical knowledge, including a rather clear statement about continental drift among the Americas, Eurasia, and Africa that was not accepted as essentially correct until the 1960s. In this geological history Buffon courageously specified distinct epochs—coincidentally or significantly (we cannot be certain) he proposed seven of them—to span a (then-vast) million years or so. This brought him into direct conflict with the Church, which by this time had more or less adopted Anglican Archbishop James Ussher's estimate of 4004 B.C. as the "official" date of creation.

Despite his occasional deistic leanings and his attempted compromise between religion and science, on these and other issues (including his proposal that Earth originated from matter freed by the glancing blow of a comet against the molten surface of our Sun) Buffon, like Galileo, was forced by dogmatic theologians to retract; his works were censured and condemned, his persuasiveness silenced during his own lifetime.

Buffon might have been influenced by a compatriot evolutionist who speculated broadly during the early eighteenth-century French Enlightenment. A remarkable work of fiction by the geologist Benoît de Maillet (1656–1738) appeared posthumously in 1748. Fearing assault by orthodox theologians—the book's subtitle gives us an idea why: *Conversations Between an Indian Philosopher and a French Missionary on the Diminution of the Sea*—he not only was careful to have his ideas published after his death but also (thinly) disguised his authorship by entitling the book *Telliamed*, which is his own name reversed. Weaving facts into a romantic Oriental background, de Maillet did a credible job of attempting to account for the gradual origin of the Sun and planets, oceans and mountains, strata containing fossil shells, the appearance of life in shallow waters, and the subsequent colonization of land by plants and animals. He fully realized that, given basic similarities in the anatomies of different animals, these animals must have metamorphosed one into another by specialization of their parts. But he fallaciously contended, like Buffon later, that such modifications occur within the short period of a single lifetime. Effectively, de Maillet's scheme amounted to a rather crude revival of Empedocles' "transmission of acquired characters," an intriguing concept made popular by another pre-Darwinian natural historian who often gets credit for it.

Jean Baptiste Lamarck (1744–1829), another in the steady stream of enlightened French naturalists of the eighteenth century, fully developed the idea of the "inheritance of acquired characters" and is often considered the founder of modern evolutionism. He is even credited with having coined the term "biology" as a sort of hybrid to denote the synthetic study of botany and zoology.

Lamarck's principal idea somewhat mimicked the speculations (on mutation, adaptation, and the struggle for existence) of Erasmus Darwin (1731–1802), Charles Darwin's physician grandfather and author of the then-influential (at least in Britain) medicophilosophical work *Zoonomia* (1794–1796). But he differed considerably from the viewpoint of the sometimes deistically inspired Buffon. In particular, Lamarck's masterwork, *Zoological Philosophy* (1809), theorized that the animal kingdom displays a fundamental unity, that mutable species vary in response to changing environmental influences, and that all life is characterized by progressive development. His was the first significant attempt to explain comprehensively biological evolution as we now understand it.

Lamarck also proposed that life's many varied and diverse species be mapped according to the now-familiar shape of a tree or bush, with life forms branching from the trunk into a network of larger and smaller stems. After more than twenty centuries, one of the last surviving features of Aristotelianism—the linear chain of beings from simple to complex life—was beginning to crumble. (Even today zoologists devote a large part of their effort toward deciphering the great phyletic tree of life; this is particularly true among modern paleoanthropologists who constantly seek to refine the multiple branches representing the evolutionary paths of our hominid ancestors of a few million years ago.)

In a pair of landmark passages, Lamarck first placed the problem of evolution into historical perspective:

> In considering the natural order of animals, the very positive gradation which exists in their structure, organization, and in the number as well as in the perfection of their faculties, is very far removed from being a new truth, because the Greeks themselves fully perceived it; but they were unable to expose the principles and the proofs of this evolution, because they lacked the knowledge necessary to establish it.

In the same paragraph, and in a spirit quite opposed to the then-popular dogma of spontaneous creation and its associated fixity of species, he made clear his preference for a bona fide evolutionary process (though he also revealed his anthropocentrism by regarding humans as perfect):

. . . with life forms branching . . .

. . . Nature, in producing successively all the species of animals, and commencing by the most imperfect or the most simple to conclude its labor in the most perfect, has gradually completed their organization; and of these animals, while spreading generally in all the habitable regions of the globe, each species has received, under the influence of environment which it has encountered, the habits which we recognize and the modifications in its parts which observation reveals in it.

Favoring the uniformitarian (or slow-change), as opposed to catastrophic (or violent-change), school of geology, Lamarck repeatedly stressed the gradual and even sluggish tenor of the events that yield change, both physically and biologically. In a statement noted both for its sweeping generalization and for its understated truism, he noted: "For all the evolution of the Earth and of living beings, nature needs but three elements—space, time, and matter." Again, he later emphasized that for "Nature to perfect and diversify animals requires merely matter, space, and time." Of course such grand statements, however correct, do not really explain much. To the extent that they imply an emphasis on long durations, these comments comprised a valid contribution at the time, but specifics are required to make a viable advance in any subject.

Lamarck's central thesis has come to be known as the law of use and disuse. Here he argued that environment does not directly (and therefore rapidly) produce changes in a life form (as both de Maillet and Buffon had essentially maintained), rather that nature engenders gradual change indirectly according to a life form's reaction toward its environment:

In every animal which has not passed the term of its development, the more frequent and sustained employment of each organ strengthens little by little this organ, develops it, increases it in size, and gives it a power proportional to the length of its employment; whereas the constant lack of use of the same organ insensibly weakens it, deteriorates it, progressively diminishes its powers, and ends by causing it to disappear.

That is the "acquired characters" part of Lamarck's evolutionary concept, after which he wrote of "inheritance": "All that has been acquired or changed in the organization of individuals during their life is preserved by generation, and transmitted to new individuals which proceed from those which have undergone these changes."

In the classical example cited in most of today's biology textbooks, Lamarckism maintains that giraffes have long necks because the necks of some of their predecessors were used extensively, even stretched, in order to reach the needed food high in the trees of the African plains. However, controlled experimental tests have since strongly suggested that these contributions are factually incorrect. And in light of Charles Darwin's later success, Lamarck's explanation for the mechanism of biological evolution is no longer tenable within the scientific community. Lamarck made no lasting contributions cosmologically.

The identity of the pre-Darwinian natural historian who impresses me most is not entirely certain. The anonymously authored and impressively perceptive *Vestiges of the Natural History of Creation* (1844) had to have been written by a naturalist of (then) liberal views and considerable knowledge of geology; many scholars suspect this person to have been the Edinburgh encyclopedist Robert Chambers (1802–1871). More than championing the role of change among life forms, this work comprises a vigorous and sustained (though by no means error-free) appeal for an evolutionary system of the whole Universe, a marvelous attempt at a grand synthesis from nebula to human, employing astronomy, biology, geology, and anthropology. In short, a genuine piece of heresy, given the stuffy Victorian environment in which it was penned.

Vestiges of . . . Creation contains, up to that time, the strongest presentation of the scientific evidence favoring cosmic evolution in its lengthy battle against spontaneous creation. The author, whom I too shall assume was Chambers, began with an analysis of the then-known Solar System, reasoned that all heavenly bodies probably contain elements of the same kind, argued that life most likely originated from inorganic matter, recognized the geological record's inference of a perennially ascending trend of fossil complexity, and concluded by contending that humans are the most recent development in the animal kingdom. Remember, all this collective insight was posited in a single volume some one hundred fifty years ago.

Chambers embraced Buffon's idea of environmental change: "Light, heat, and the chemical constitution of the atmosphere may have been the immediate prompting cause of all those advances from species to species. . . ." At another point he cited the progressionist principles of Aristotle, suggesting that life results from a first impulse imparted by God, after which through various grades of organization all life forms advance over eons from the lowest to the highest plants and animals. Since this initial (divine) impulse might produce types ill suited to their environment, Chambers added a second (natural) impulse that purportedly adapts organic structures according to their environments, food, and habitats.

As noted above, his ideas of evolutionary continuity were not limited to life forms. Consider his conviction on the connections among various bodies in space: ". . . as the planets are parts of solar systems, so are solar systems parts of what may be called astral systems—that is, systems composed of a multitude of stars, bearing a certain relation to each other." No doubt the pioneering research by the eighteenth-century British astronomer William Herschel had already addressed the larger swarm of stars that we now term the Milky Way, even likening the variety of objects observed in the heavens to a luxuriant vegetable garden. But Chambers' insight lies in his recognition of the linking gradations among stellar systems much as had been earlier proposed for life forms on Earth.

Though nearly a century was to pass before the formal inauguration of the field of stellar evolution, Chambers further illustrated the interrelationships among stellar systems in this utterly prophetic and remarkable passage:

> The two Herschels have in succession made some other remarkable observations on the regions of space. They have found within the limits of our astral system, and generally in its outer fields, a great number of objects which, from their foggy appearance, are called nebulae; some of vast extent and irregular figure, as that in the sword of Orion, which is visible to the naked eye, others of shape more defined; others, again, in which small bright nuclei appear here and there over the surface. Between this last form and another class of objects, which appear as clusters of nuclei with nebulous matter around each nucleus, there is but a step in what appears a chain of related things. Then, again, our astral space shows what are called

95

nebulous stars—namely, luminous spherical objects, bright in the center and dull towards the extremities. These appear to be only an advanced condition of the class of objects above described. Finally, nebulous stars exist in every stage of concentration, down to that state in which we see only a common star with a slight bur around it. It may be presumed that all these are but stages in a progress, just as if, seeing a child, a boy, a youth, a middle-aged, and an old man together, we might presume that the whole were only variations of one being. Are we to suppose that we have got a glimpse of the process through which a sun goes between its original condition, as a mass of diffused nebulous matter, and its full-formed state as a compact body?

Later Chambers continued his cosmic-evolutionary odyssey:

. . . seeing in our astral system many thousands of worlds in all stages of formation, from the most rudimental to that immediately preceding the present condition to those we deem perfect, it is unavoidable to conclude that all the perfect have gone through the various stages which we see in the rudimental. This leads us at once to the conclusion that the whole of our firmament was at one time a diffused mass of nebulous matter, extending through the space which it still occupies. So also, of course, must have been the other astral systems. Indeed, we must presume the whole to have been originally in one connected mass, the astral systems being only the first division into parts, and solar systems the second.

Not only did he articulate, with what we now judge to be much qualitative validity, an origin for our Solar System, but Chambers also briefly recounted earlier epochs of the Universe. Note how he stressed the need for change:

The nebulous matter of space, previously to the formation of stellar and planetary bodies, must have been a universal Fire Mist, an idea which we can scarcely comprehend, though the reasons for arriving at it seem irresistable. The formation of systems out of this matter implies a change of some kind with regard to the condition of the heat.

96

He went on to admit that his scenario is nonquantitative, a posture frankly not surprising given that his arguments predated the formal development of the subject of thermodynamics (to be discussed in the next chapter). Said Chambers, again perceptively for the time: "We do not know enough of the laws of heat to enable us to surmise how the necessary change in this respect was brought about. . . ."

In much of this, Chambers was careful to point out that he regarded his work as contributing to the greater glory of God, all the while vehemently repudiating atheism. To him, a declared theist, the Creator displays quintessential dignity by working through natural laws.

Still, as Chambers perhaps suspected, his *Vestiges of . . . Creation* won only scathing reviews, each claiming it full of false science and a lack of religious faith. After all, the Church had its doctrine to protect, the contented Victorian state didn't want its status quo disturbed, and the many practitioners of (Aristotelian) science had their reputations to guard. Proclaiming Chambers' work "prophetic of infidel times and unsound of our general education," the *North British Review* announced that it "poisoned the fountains of science and sapped the foundations of religion." In yet another book review which ran a total of four hundred(!) pages, a leading Cambridge don of science objected violently: "The world cannot bear to be turned upside down and we are ready to wage an internecine war with any violation of our modest principles and social manners. . . ." With advertisements like these, even in the mid-nineteenth century, the book quickly became a best-seller; ten editions were sold out in as many years.

As a self-acknowledged amateur of the history of science, I can well remember my equally strong reaction when I serendipitously happened across an original edition of *Vestiges of . . . Creation* a few years ago. Nearly locked up after hours in the bowels of Harvard's cavernous Widener Library, I read with studied amazement and a rising sense of excitement, but my perception was distinctly opposite to that of my scientific counterparts of a century and a half ago (though I acknowledge the benefit of hindsight). To be sure, my "discovery" that day made a lasting impact on me, not least because here was an erudite work that addressed the forest when the scientific establishment saw only the trees, but also because Chambers felt a need to make his contribution anonymously lest he be ostracized as much by the hierarchy of the scientific world as by the religious community. As I have said in another context, nothing is immutable, not even prevailing scientific dogma. I took pleasure in noting an unsigned inscription penciled on the inside flap of the yellowing, ribbon-bound and tattered volume, obviously donated from a private estate and apparently last borrowed in 1871: "This book when first published and where father bought it was considered a very bad book. And created a sensation in the religious world." And even with the Ice Age expert (though avowed creationist) Louis Agassiz then in command of Harvard science, someone had felt a need to paste on the titlepage a barely discernible newspaper clipping apparently from a review by H. T. Buckle, a noted mid-nineteenth-century English historian, which I cannot resist quoting:

> Every new truth which has ever been propounded has, for a time, caused mischief; it has produced discomfort, and often unhappiness; sometimes by disturbing social or religious arrangements, and sometimes merely by the disruption of old and cherished association of thoughts. It is only after a certain interval, and when the framework of affairs has adjusted itself to the new truth, that its good efforts preponderate.

> *Darwin and Mendel laid on man the chains*
> *That bind him to the past. Ancestral gains,*
> *So pleasant for a spell, of late bizarre,*
> *Suggest that where he was is where we are.*
> *—David McCord*

By the mid-nineteenth century, even with the numerous facts that had accumulated about it, the idea of evolution was in eclipse. States opposed it, the Church denied it, and even most leaders of the scientific establishment argued against it. Many of the times' most influential naturalists—notably Georges Cuvier in France; Adam Sedgwick, John Henslow, and Richard Owen in Britain; and Cuvier's student Louis Agassiz in America—espoused the fixity of species and thus were antievolutionists. Cuvier, for instance, repudiated the possibility of evolution, first, because the fossil record displayed no intermediate stages (known to him) connecting one group with another and, second, because the tombs of pharaohs, revealed by archeologists who accompanied Napoleon's military expedition to Egypt, contained mummified humans and animals virtually identical to those then alive, "proving" that if human anatomy had not changed in six thousand years then the living world must be naturally steadfast. These and other scientific leaders of the time preferred to invoke geological catastrophism (such as global floods, including the Noachian Deluge) to explain the undeniable fossil evidence for the extinction of species, to be replaced with new species by means of miraculous spontaneous creation. Charles Lyell (1797–1875), the British historical geologist, was an important exception, for his materialistic sixteen-hundred-page tome, *Principles of Geology* (1830–33) seemingly paved the way for gradualistic Darwinism. Lyell's uniformitarianism asserted geological change to be dynamic yet slow and continuous, although he personally resisted evolutionary explanations for the origins of new plant and animal species.

As in the few centuries subsequent to Empedoclean times, biological evolutionism had once again become almost wholly subordinate to a lingering yet pervasive theism—this despite the deterministic successes of the physical sciences having made belief in literal biblical interpretations increasingly unfashionable. Thus, although Darwin later acknowledged that Chambers' *Vestiges of . . . Creation* had served well in "removing prejudice and thereby preparing the ground for the reception of analogous views," the stage was set for the gravest challenge to the traditionally static Judeo-Christian worldview. This challenge was the publication, in 1859, of one of the most brilliant scientific works of all time, *On the Origin of Species.*

In a brief survey such as this, I find it impossible to treat properly the work and influence of Charles Darwin (1809–1882). Indeed, I judge it equally difficult to say much about him or his accomplishments that has not been better said elsewhere. What I do suggest, as noted in the introduction to this chapter, is that Darwin probably owed more to past efforts than is generally believed (or than he publicly acknowledged). The idea of evolution, though then a minority view even among naturalists, was clearly "in the air" by the mid-nineteenth century. That different life forms shared a common ancestry, a concept now commonly accepted among research scientists worldwide, had been proposed by numerous of Darwin's predecessors even as far back as antiquity. His signal contribution was the vast quantities of data that made the idea scientifically and popularly acceptable. What's more, in a stroke of genius Darwin advanced beyond where others had trod before; he proposed a *cause* of evolution—namely, a convincing, testable mechanism describing *how* evolution works. And not least, Darwin knew well how to communicate, for although he apparently loathed writing—"A naturalist's life would be a happy one if he had only to observe and never to write"—his clarity and literary style comprise a model of scientific exposition at its best.

In enunciating his theory, Darwin combined his own observations of nature and a deeply ingrained belief in the scientific (inductive) method of approach. Marshaling massive amounts of empirical evidence (especially during his five-year circumnavigation aboard HMS *Beagle*), whereas his predecessors had sought little, Darwin was perhaps the first evolutionist to fully utilize genuine Baconian principles. By observing, experimenting, and testing, he was able, in a single lifetime, to leap over the speculations of centuries, even millennia.

Darwin's genius was his theory of natural selection: Since all plants and animals multiply faster than nature can provide nourishment for them, many in each generation will perish before reaching the age of reproduction. (This Darwin garnered from a chance though beneficial reading of English economist Thomas Malthus' classic work *An Essay on the Principle of Population* [1798].) Accordingly the living world is characterized by an inevitable struggle for existence, ensuring that only the favored (some claim the "fittest") survive and avoid extinction. That is to say, in a given habitat members of the same species compete for survival; those better adapted to the environment have a greater opportunity not only to survive but also to reproduce and thereby perpetuate their kind within the limited economy of nature. Thus, random variations (now termed "genetic mutations") in physical stature or behavioral pattern that confer a survival advantage will, more likely than not, predominate in adults within each generation. Given enough time, said Darwin, the responses of organisms to chance variations and environmental change could account for the whole developmental lineage from single cells to human beings. These two factors—

genetic and environmental alteration—working in consort with natural selection comprise the causal mechanism to account for the origin of species. And as noted in my previous *Cosmic Dawn,* among many other works, the principle of natural selection can be observed and analyzed with considerable precision, provided the environmental conditions and variables are strictly controlled.

The "chanciness" of Darwinian evolution caused thinking humans to become less mechanistic. The deterministic view of our world had clearly softened. Today we recognize that natural selection occurs according to the laws of molecular dynamics and is not strictly the result of chance. Chance indeed dictates which mutations occur, and in what sequence. But it is a matter of predictable necessity that mutations will occur and that selection will take place among them. This combination of chance and necessity suffices to explain the tendency, inherent in biological evolution, for adaptation and improvement over the course of time.

Where did Darwin stand regarding a Grand Designer in nature? In correspondence reproduced in *Life and Letters of Charles Darwin* (an 1887 collection of his papers compiled by his son, Francis), Darwin anticipated our contemporary views on the role of purpose or teleology, saying, "I cannot think the world, as we see it, is [wholly] the result of chance; and yet I cannot look at each separate thing as the result of Design." Even more clearly, in a later passage of the same volume this agnostic (if not tacit atheist) remarked, "I must think that it is illogical to suppose that the variations, which natural selection preserves for the good of any being, have been designed."

Perhaps no one has yet summed up Darwin's vision better than he himself in the final, eloquent paragraph of *The Origin:* "There is grandeur in this view of life, with its several powers, having been originally breathed into a few forms or into one; and that, whilst this planet has gone cycling on according to the first law of gravity, from so simple a beginning endless forms most beautiful and most wonderful have been, and are being, evolved."

Charles Darwin also presciently envisioned today's origin of life experiments—". . . if we could conceive in some warm little pond, with all sorts of ammonia and phosphoric salts, light, heat, electricity, etc., present, that a protein compound was chemically formed ready to undergo still more complex changes . . ."—but he confined his writings to the biological realm, speculating neither about larger cosmic matter nor about the import his ideas might have on philosophical and theological issues.

In succeeding years, indeed, throughout the late last century and to the present, much learned discussion and debate have defended, expanded, and disseminated Darwin's scientific theory of biological evolution. The English naturalist Alfred Wallace (1823–1913) independently proposed the principle of natural selection and early on defended Darwinism, but his evolutionary interpretation later in life became more spiritual, especially his thesis that the presence of humans on Earth can be explained only by invoking supernaturalism. The British surgeon T. H. Huxley (1825–1895) asserted in his *Evidence as to Man's Place in Nature* (1863) that humans share a common ancestry with the great apes, thus substantially anticipating Darwin's other major work, *The Descent of Man* (1871). The English agnostic Herbert Spencer (1820–1903), in a comprehensive ten-volume set entitled *Synthetic Philosophy* (1860–96), attempted to broaden the salient features of Darwinism to incorporate human society, the result being a controversial "Social Darwinism," integrating aspects of biology, psychology, sociology, and ethics. The Augustinian monk Gregor Mendel (1822–1884) fathered the science of mathematical genetics by means of heredity experiments involving the hybridization of garden pea plants, thus laying the foundation for the modern molecular basis of biological evolution. An international host of twentieth-century biologists (including Theodosius Dobzhansky, J. B. S. Haldane, Ernst Mayr, and George Gaylord Simpson) synthesized classical Darwinism with Mendelian genetics to produce the currently accepted and variously termed Neo-Darwinism or the Modern Synthesis. The American paleontologists Niles Eldredge and Stephen Jay Gould have recently stressed that biological evolution is not as gradualistic as Darwin once proposed (though their work in no way repudiates the essence of Neo-Darwinism), maintaining instead that life forms speciate during periods of rapid change amidst long intervals of stasis (a proposal called "punctuated equilibrium"), much as geological drift among tectonic plates often yields sudden, discontinuous change (during earthquakes and volcanic eruptions) despite years of inactivity. And largely on the basis of observations of ants and bees, the contemporary U.S. sociobiologist E. O. Wilson has pioneered a synthesis of the social and biological sciences, in the process suggesting that human behavior is substantially affected by our genes as well as by the environment in which we are embedded.

> Atom from atom yawns as far
> As moon from earth, or star from star.
> —Ralph Waldo Emerson

The early parts of the twentieth century saw some notable efforts to broaden the concept of evolution (though not necessarily the principle of natural selection) to include the physical sciences. These efforts, all for the most part qualitative, were varied and interdisciplinary.* The pathetically disposed German antichrist Friedrich Nietzsche (1844–1900), influenced strongly by the pre-Socratic Heraclitus, philosophically urged an infinite Universe consisting of physical energy that ceaselessly changes in an "eternally recurring" (or cyclical) manner. The German atheist Ernst Haeckel (1834–1919) advocated a (still generally sound) evolutionary cosmology wherein a universal "law of substance" was theorized as the fundamental fact of reality and which specified the motion of matter and energy as the source of all things, including the innumerable objects scattered across the cosmos, humankind itself, and human self-consciousness. The founder of pragmatism in America, Charles Peirce (1839–1914), influenced somewhat by the New England transcendentalists and especially by the Lamarckian doctrine of acquired inheritance, taught that a practical philosophy must rigorously incorporate the theory of evolution as applicable to the Universe as a whole—a prescient idea that had the cosmos evolving from a nascent state of chaotic homogeneity to a future state of ordered heterogeneity and maximum beauty. A relatively unknown Serbian eclectic, Bozidar Knezevic (1862–1905), posited quite plausibly (even today) a cosmos that evolves through three major phases— inorganic, organic, and psychic—after which it eventually devolves back into an ultimately unconscious chaos (closely resembling what

---

*A few other relatively modern scholars who embraced the concept of change, including Teilhard and Bergson, are discussed more appropriately in the context of Chapter 5, for their ideas have implications for the Life Era.

we now call a "closed" Universe). And the British-U.S. logician Alfred North Whitehead (1861–1947) formulated an "organic philosophy" wherein the many varied and complex components of the world are reasoned to be unified not only on the ground of natural science but by aesthetic, moral, and religious experience as well. Foremost among these writings (at least as regards the broader realm of universal change) are those of the former dean of American astronomers, the pantheist Harlow Shapley (1885–1972), whom I regard as the modern father of cosmic evolution.

In a remarkable book published in 1930 and aptly entitled *Flights from Chaos*, Shapley surveyed and classified all material systems from atoms to galaxies. The book's preface succinctly outlined his larger view of change in the cosmos: ". . . interstellar and intergalactic space, which is traversed at all points with the radiation of all stars, is found to be of increasing significance because such regions are the graves of expiring stars and may contain the cosmic soil from which new suns and galaxies arise." The book is absolutely prophetic of the research now under way in cosmic evolution, though scientists of a half century ago doubtless accorded it little dignity because of its "popularized" nature. Nor did his work quantify a mechanism for universal change, a common denominator to underlie the evolution of material systems.

Shapley's avowed goal was to untangle the order in the Universe from the apparent chaos of confusing motions, varied types, complex radiation, and irregular organizations. From the minutest atomic structures to the largest stellar groups, Shapley argued that prudent inspection and analysis transform evidences of apparent chaos into a high degree of order. "In the animal and vegetable kingdoms, the biologist has made progress towards order, and his numerous classifications have provided some of the materials needed for contemplation of the origin, growth, and destiny of organic forms. The astronomer, also, has made some advance toward untangling the thread of sidereal structure." He went on to list his early-twentieth-century research objectives—namely, to classify star clusters, interstellar clouds, and remote galaxies—with an eye toward unraveling an overall "view of organization in the cosmos" and all as part of his general studies of taxonomic problems in the field of what he called "cosmography." (According to Shapley, "cosmography is to the cosmos what geography is to the Earth," thus demonstrating further that his objective was classification more than explanation.)

Everywhere Shapley envisaged organization—both microscopically and celestially. Often through classification alone he perceived order amidst chaos. In his *Flights from Chaos* as well as his later, principal exposition of cosmic evolution, *Of Stars and Men* (1958)—both works now somewhat dated—he considered the structure of atomic and molecular systems, using them as stepping-stones to higher assemblages. In turn, "molecules grade imperceptibly into crystals and colloids; and these two forms of molecular structures appear more or less definitely as units in higher aggregates." Noting that no sharp demarcation divides the simplest molecule of hydrogen from the largest terrestrial organism, Shapley proposed a simple classification of material systems, spanning the submicroscopic and microscopic worlds: corpuscles (which we now call "elementary particles"); atoms; molecules; molecular systems (i.e., crystals as well as colloids, which are finely divided solid particles suspended in a liquid); inorganic aggregates (e.g., minerals, meteorites, clouds); and organic substances (e.g., organisms, colonies, humans, etc.).

These inorganic and organic substances represent a transitional group marking the point where the organizing forces cease to be electromagnetic and become gravitational. In addition to a change in the nature of the controlling force, this group represents a transition point for dimension. According to Shapley, this latter group "marks a crossing from the microscopic to the macroscopic, from the essentially physico-chemical field to the realm that is predominantly astronomical; and at the crossing, mineralogy, geology, and biology appear."

Shapley classified gravitational systems according to increasing mass and dimension—that is, his scale depended on structure, not on age. This is the way that he perceived macroscopic order amongst an apparently chaotic confusion in nature. For example, he noted that some ninety percent of all stars in our Milky Way fall into one of two categories: galactic clusters, often containing dozens or sometimes hundreds of young stars, and globular clusters, usually housing hundreds of thousands of mostly old stars.

Incidentally, the globular star clusters are the very same stellar associations that Shapley used to discover the size and scale of our

Milky Way in the 1920s. In particular, he found that the globulars are centered on a locality some thirty thousand light-years from the Sun, proving that our Solar System is not at the previously believed special position of the center of the Galaxy—or at the center of the Universe for that matter, for even in the early part of the present century the Galaxy was essentially taken to equal the Universe. His work on the stellar census was an enormous step forward in the human understanding of our place in the cosmos, for it amounted to no less than an obituary for anthropocentrism in our description of the Universe. No small breakthrough, Shapley's research on the positions of the star clusters did for the Sun what Copernicus' study on the positions of the planets had previously done for the Earth. Though few people realize how recent an accomplishment this was, scarcely more than fifty years ago Shapley endowed our Sun with the same mediocrity that Copernicus (and Aristarchus before him) had argued for the Earth nearly five hundred years ago. Now both Sun and Earth were removed forevermore from any "center" or privileged position in the Universe—hence, the origin of Shapley's quip about our living in the "galactic suburbs," as noted in Chapter 1.

Shapley pressed on to include in his classification the then recently discovered remote galaxies. Furthermore, he discussed multiple galaxies linked by gravity (which we now call "galaxy clusters"), as well as "higher combinations of sidereal systems," including an intriguing but unspecified class that he termed "transcendent superorganizations." Continuing, he stated, "As the Solar System is to our Galaxy with its millions of stars, so is our Galaxy to the higher system of systems."

All in all, Shapley assembled his total classification of material systems in much the same way that a student of cosmic evolution would do today. In order of increasing size and mass, we have elementary particles, atoms, molecules, molecular systems, organic and inorganic aggregates, meteoroids, moons, planets, stars, star clusters, galaxies, galaxy clusters, Universe. And like Shapley, we would stress that though tentative in parts and not necessarily complete, this classification (of *systems*, not objects per se) does represent progress toward order but perhaps not a complete escape from chaos. Shapley summed up his position smartly, cautioning, "At best, it is a working classification, designed to give perspective and to guide further inquiry into the relationships of types of organization."

Later, in his 1967 memoirs, *Beyond the Observatory,* Shapley went beyond mere classification, alluding (though only qualitatively) to an evolutionary link among all things: ". . . nothing seems to be more important philosophically than the revelation that the evolutionary drive, which has in recent years swept over the whole field of biology, also includes in its sweep the evolution of galaxies and stars, and comets, and atoms, and indeed all things material." In the same volume he became the quintessential cosmic evolutionist, saying:

> The word evolution is commonly associated with evolving plants and animals, nothing else. It suggests changes in the biological world. . . . Yet for a century scientists have been aware of evolution beyond the biological kingdom. We have seen that volcanic action and the oxydizing of lava rocks indicate that this planet's surface still changes with time; it evolves, and our atmosphere does also. The variety among stars suggests stellar evolution. For half a century we have realized that the fact that our Sun is shining is evidence that it is steadily losing mass-energy . . . and therefore that it is evolving. Similarly, therefore, star shine must mean stellar evolution, a much mightier operation than we can muster among the animals and plants and noncellular organisms on the Earth's surface. It is only a short step from stellar evolution to galaxy evolution. Going further, we see that the discovery of the expanding Universe indicates that growth, change, evolution affect also the future of the galaxy of galaxies.

Intriguingly, Shapley listed in his cosmography an ultimate class, saying simply that "we should allow for systems beyond those now conceived" and adding, "Scientific pronouncements concerning unsurpassable limits in dimensions and masses or ultimates in organizations are likely to be mere dogma." Given his definition that "In our present usage, the Universe includes every material thing we know . . . ," I've often wondered why he left room for an additional, unspecified class seemingly beyond that of the Universe. What did he mean? Was the foremost American astronomer of the time also a teleologist?

Perhaps, but perhaps not, for throughout much of his writings Shapley spoke of the Universe's being structured by a framework of four basic entities—space, time, matter, and energy—as well as a fifth, potentially superior (i.e., more basic) property that he variously termed "direction, form, drive, consciousness," and, more often, "cosmic evolution, or natural logic." In *Of Stars and Men*, he speculated that this fifth, master "property of the material world [might be] essential to make the Universe go"—perhaps an indispensable "something that would make click a Universe of stars, organisms, and natural laws . . ."—but just as quickly in direct reference to the concept of God he added that "we should not be too hasty in such a deep and critical matter."

The only time I met Harlow Shapley, I took advantage of the opportunity to pick his brain. I was the impressionable young Harvard assistant professor with my newly minted doctorate; he the professor emeritus, feeble, wobbly, nearly ninety years of age. Regardless, as the modern father of cosmic evolution he seemed to me to embody the ultimate universal class himself. Upon discussing cosmic evolution, he became as sharp as tacks, much as though he had summoned some inner strength (a Force perhaps!) when questioned about his grand scheme of nature. His enthusiasm was especially evident when I told him I had intentions of trying to resurrect the core of his legendary Harvard course that had lain dormant for decades (as I later helped to do until, after it had had an abnormally successful decade-long run, Arts-and-Science officials declared the course "too general for general education"). Walking with him from the faculty club that warm spring day in the early 1970s, I put my query to him: Why a supra-Universe class? He stopped, looked me directly in the eye, and repeated the only justification he had offered in his *Flights from Chaos:* ". . . partly as a matter of safety, partly as a challenge." He died shortly thereafter.

CHAPTER 3

Physics of Change

THE FUNDAMENTAL LAWS OF THERMODYNAMICS GUIDE THE CHANGES AMONG ALL THINGS IN THE UNIVERSE

COSMIC EVOLUTION, AS WE UNDERSTAND IT TODAY, is governed largely by the laws of physics, in particular those of thermodynamics. That's because, of all the known principles of nature, thermodynamics has the most to say about the concept of change. Literally "thermodynamics" means "movement of heat"; for our purposes in this book (and in keeping with the wider Greek connotation of motion as change), a more insightful translation would be "change of energy."

> Matter whose movement moves us all
> Moves to its random funeral,
> And Gresham's law that fits the purse
> Seems to fit the universe.
> Against the drift what form can move?
> (The God of order is called Love.)
> —Herbert Spencer

109

The first law of thermodynamics is a conservation principle. It states that all energy in the Universe is constant—that is, the sum of all energy is fixed, has been fixed since the beginning of time, and will remain so until the end of time. Even so, energy can appear in various forms, for example, heat, light, gravitation, invisible radiation, kinetic energy, mechanical work, chemical potential, nuclear energy, and so forth; matter itself is a form of energy. (The eighteenth-century poet William Blake referred to energy as "eternal delight.") Furthermore, the many varied forms of energy can be interchanged, including matter transforming into energy and conversely, as Einstein taught us, a fact made clearly evident by the development of atomic weapons. But the *sum* of all forms of energy remains constant for all time. In short, the first law of thermodynamics decrees that energy itself can be neither created nor destroyed, though its many forms can change.

If the first law were the totality of thermodynamics, we could interchange energy among its varied forms (including matter) without limit. Alas, there exists another basic principle of thermodynamics, a second law, which is more subtle than the first. Initially expressed scientifically around 1850, independently by physicists Lord Kelvin of Britain and Rudolf Clausius of Germany, the second law specifies the way in which change occurs in a quantity called "available energy," which is also on occasion variously termed "usable energy," "free energy," or "potential energy." This law's essence stipulates that a price is paid each time energy changes from one form to another. The price paid (to nature) is a loss in the amount of available energy capable of performing work of some kind in the future. Physicists have a term characterizing this decrease in available energy; we call it "entropy." Clausius himself, in 1865, explained the origin of this peculiar name: "This term is based on the Greek word *tropae*, meaning 'transformation.' I have deliberately made the structure of this word analogous to that of 'energy,' because the two quantities described by these terms [energy and entropy] are so closely related in physics that the parallel seems useful to me here."

Entropy is a measure of the amount of energy no longer capable of conversion into useful work. Numerically entropy is expressed as a fraction; it equals the amount of heat exchanged from one part of the Universe to another, divided by the temperature at which the change occurs. It is also a measure of the disorder (or randomness) of a system, whether that system be something as small as a crystal made of molecules or as large as a Universe of galaxies. Since entropy plays such a central role in many of the ideas presented in parts of this book, it is worth discussing its salient features further.

In 1824 a young French army officer, Sadi Carnot, sought to understand the rudiments of an ordinary steam engine. He discovered that such engines work because of a temperature difference; part of the engine is cold while another part is hot. Indeed, for any engine to convert energy into useful work, there must be a temperature differential (or thermal gradient). Work then occurs when energy (heat in the case of a steam engine) flows from a body of higher temperature to one of lower temperature—in other words, from a higher energy state to a lower energy state. Carnot articulated his finding succinctly in his memoirs, *Reflections on the Motive Power of Fire:* "For a heat engine operating in cycles to perform mechanical work, we must use two bodies of different temperatures." Equally important, Carnot discovered that each time energy flows from one state to another, less energy is available to perform work the next time around. Energy is not lost, just rendered unavailable for useful work.

Thus, if any isolated physical system—one closed to the outside world—is divided into two parts, energy can flow from one part to the other, but the total energy of the system cannot be increased or decreased. This, in essence, is a statement of the first law of thermodynamics. Accompanying this internal flow of energy will also be a flow (or change) of entropy. Only for ideal (so-called reversible) processes will the resulting change in the entropy of the system be zero. For realistic (i.e., irreversible) processes, there must be an increase in the *total* entropy, for with each process less energy is available for conversion to useful work. This in turn is a statement of the second law of thermodynamics.

What is the relative importance of the two laws? Sir Arthur Eddington once expressed it rather directly: "If a young scientist comes to us who doubts the first law, we give him patient hearing. But if he doubts the second law, we suggest tactfully that he would be perhaps happier in another profession." On another occasion Eddington again held high the primacy of the second law:

> The law that entropy always increases—the second law of thermodynamics—holds, I think, the supreme position among the

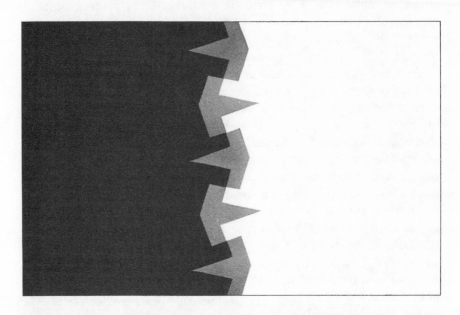

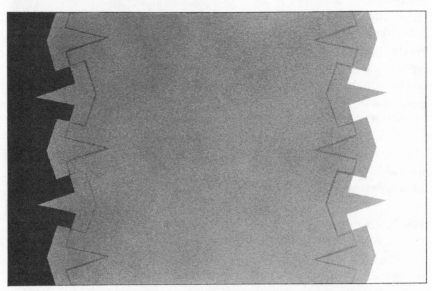

. . . each time energy flows from one state to another, less energy is available to perform work the next time around.

112

laws of Nature. If someone points out to you that your pet theory of the Universe is in disagreement with Maxwell's equations—then so much the worse for Maxwell's equations. If it is found to be contradicted by observation—well, these experimentalists do bungle things sometimes. But if your theory is found to be against the second law of thermodynamics I can give you no hope; there is nothing for it but to collapse in deepest humiliation.

A familiar example provides a particularly simple illustration of the above arguments. When an iron bar is heated at one end, the other end will eventually warm until the temperature of the whole bar becomes equal. This is known to anyone who has left for too long a metal stirring spoon in a pot of hot soup or a poker in a fireplace. The reverse phenomenon—namely, a uniformly warm iron bar's suddenly becoming hot at one end and cold at the other—has never been observed. Likewise, when a container partitioned into two compartments, one filled with gas and the other empty, has its partition removed, the gas will expand into the empty compartment until both parts are equally full. The reverse process—namely, the dispersed gas's abruptly congregating in only one half of the container—has also never been observed. (Can you imagine the steam released from a whistling teakettle suddenly heading back into the kettle!) All such natural (realistic) events proceed in one direction only. Nature forbids (or at least demands a penalty for) their reversal, which is why real events are described as irreversible or asymmetric: Hot objects cool, but cool objects do not spontaneously become hot; a bouncing ball comes to rest, but a stationary ball does not spontaneously begin to bounce.

Consider yet another familiar example of irreversible energy change—namely, the case of water falling over a dam into a basin below. This is a mechanical (or gravitational) analog of the thermal and chemical cases just discussed and is the kind of energy that first powered our country's mills. As a boy I grew up in the "spindle city" of Lowell, Massachusetts, where water from the Merrimack River was once diverted into an intricate system of canals built by early-nineteenth-century immigrants to provide the primary source of energy to turn huge waterwheels mechanically. Later, starting around the turn of the century and continuing to the present, part of this same system powered spinning turbines that generated elec-

113

tricity for much of the city. Many of these man-made canals employ locks, or artificial dams, to hold the water temporarily at certain levels, after which the water can experience a drop in level and thus power the weaving machines that once manufactured textiles. Having traversed the canal system and reached the river basin below, however, the water can no longer perform work. Still water in a flat basin or lake cannot be used to turn even the smallest turbine. Nor can the basin's water be returned to the top of the dam without work being performed by some outside agent (such as a water pump), for this realistic example is an irreversible process. (Of course, as agents of change, we could use just such a pump, but this artificial process is decidedly uneconomical; more energy is required than is produced.) In short, water in a reservoir above a dam possesses some available or free energy—something we often call "gravitational potential energy," which is yet another form of energy that can be released as the water falls to a lower level closer to the center of the Earth. Provided that an energy difference exists (in this case of a gravitational nature owing to different heights of the water), then useful work can occur. In the basin below, the water still harbors energy, but it's unavailable for useful work since the energy of the water in a flat basin is everywhere the same. Furthermore, each time more energy is made unavailable for work (in this case by falling over the dam), entropy increases. As America's first industrial spy (who "photographically" memorized textile-manufacturing techniques during a visit around 1800 to the heart of the Industrial Revolution at Manchester, England), Francis Cabot Lowell and his Boston entrepreneurial backers were essentially performing a controlled experiment in gravitational contraction of the Earth—a much more efficient way to produce energy than in a chemical reaction or nuclear power plant.

So, the first law states that energy can be neither created nor destroyed, while the second law stipulates that energy can change in only one way: irreversibly toward a dissipated (randomized) state of increased entropy. Nature is said to be intrinsically asymmetric.

Now, let us again restrict ourselves to a closed system, one isolated from its surrounding environment and into which no new energy (or matter) flows. In such a system, energy states always tend

114

to even out, that is, achieve an equilibrium. In all the examples cited above—heat spreading along a metal bar, gas diffusing within a bounded container, water flowing into a river basin—equilibrium is the end product. What do we mean by equilibrium? Consider the following operational cases.

Everyone intuitively understands the nature of an equilibrium. A balance scale, for instance, maintains equilibrium when both its pans hold equal weights. If we tap the beam, the scale will oscillate, as we have imparted to the system some potential energy which converts into kinetic energy and then back into potential energy and so on. The total energy of the system remains constant, and the scale would oscillate indefinitely if it were not for friction. Realistically, however, friction will eventually cause the oscillations to diminish, and the energy contained in the (directed) oscillatory motions will become changed into the (undirected) heat motions of the individual atoms comprising the scale—all the while increasing its entropy while striving to achieve that state of equilibrium typified by the motionless scale in complete balance. Likewise, a pendulum achieves momentary equilibrium each time its arc is centered in the middle of its swing. Even the "balance of power" characterizing today's international politics is an equilibrium of sorts.

Burning wood in a stove or fireplace is yet another example of irreversible change toward equilibrium. The fire causes heat (i.e., infrared energy) to radiate and thus warm a room, as the second law of thermodynamics stipulates that heat always flows from the hotter body (the stove) to the colder body (the surrounding air). Eventually, though, the wood will have burned completely and its remaining ashes will have reached the same temperature as the air in the room. Whereas previously there was a distinct difference in energy states with the fire burning, day-old ashes exhibit no differences in energy states and have thus achieved an equilibrium.

Like water in a flat basin or a pendulum at rest, spent wood is no longer capable of performing useful work on its own. Of course, additional water could be hoisted to the top of the dam and additional wood could be placed into the hearth, but that would involve the use of a new source of available or free energy and thus would violate our restriction of a closed system.

The equilibrium state achieved within a closed system is

therefore a condition of maximum entropy—a stable state where energy can no longer freely perform useful work. In short, equilibrium is characterized by an absolute minimum of free energy and a consequent maximization of entropy. Interestingly enough, only in equilibrium can we not distinguish past from future, for in such a state there is no direction of change; the concept of irreversibility itself ceases at equilibrium.

Recall that I earlier equated entropy with randomness; the greater the randomness or disorder, the greater the entropy. To see this, we can regard the observable macroscopic properties of any system as the sum of a great many microscopic properties—perhaps the energies of electrons in an atom, or the vibrations of atoms in a molecule, or the motions of molecules in a gas. Any system having great randomness—that is, high entropy—is one having these and other microscopic properties arranged in a great many different ways. Conversely, a system of low entropy has only a few possible arrangements of its microscopic properties.

A familiar example of entropy, randomness, disorder, and chaos is a cluttered room. There a given book or pencil might be almost anywhere—on a shelf, windowsill, chair, bed, floor, wherever. In this way we see how disorder is associated with a large number of different possibilities for where any specific object could be.

Another telling illustration is the state of any university library at the end of each semester. Since I hold high the need for clear communication in science, I require each of my students, in every course I teach, to write a term paper. This inevitably leads to considerable chaos in the library, for it is a common phenomenon of college life that students (and faculty, too) cannot seem to reshelve library books properly according to their call numbers. Thus, as students research their topics toward the close of the term, the books are constantly coming off the shelves but more often that not are incorrectly replaced. And of course, if a library book is improperly shelved, it is nearly as good as lost. When the term ends, I scan the library from a distance (macroscopically) and all seems fine: Few books are missing, and the shelves are fully populated. However, upon more careful examination (microscopically), one finds that chaos rules, and only a considerable amount of work (from a kindly librarian) can regenerate order among the volumes. The natural use of a library causes order to break down into chaos; if not periodically checked, entropy will tend toward a maximum in any system.

As yet another example, imagine, if you can, an utterly frozen crystal at absolute zero temperature. Such a system can have only one possible order for its many molecular parts; thus, its entropy is zero. On the other hand, a gas at ordinary (room) temperature displays high randomness in the distribution of its atoms and molecules; it is said to have high entropy.

Accordingly, we can also view thermodynamics' second law in another, more profound way. In addition to the notion of energy flowing from available to unavailable states or from high to low concentrations (temperatures), we can alternatively regard energy in an isolated, closed system as flowing from ordered to disordered (randomized) states. Minimum entropy states, where energy concentrations are high and available energy is maximized, are considered ordered states. By contrast, maximum entropy states, where available energy is not concentrated but rather is dissipated and diffused, are considered disordered states.

Thermodynamically, then, order or organization is measured according to the number of possible arrangements of a system's microscopic parts. If a system can be described in terms of only a few such arrangements, we say that the system is very orderly. Conversely, if these parts enjoy great freedom in their arrangement, so that they can be described only in terms of many possible arrangements, then we say that the system is highly random or disorderly. And we know that the degree of randomness determines the entropy.

Here's the way the second law works in nature. Left to itself, nothing will proceed spontaneously toward a more ordered state. Housework is a familiar example. Left unattended, houses grow more disorderly; lawns become underbrush, kitchens greasy, roofs leaky. Even human beings who don't eat will gradually become less ordered—and die. All things, when left alone, eventually decay into chaotic, randomized, and less ordered states.

Of course, by expending energy, order can be reachieved. Some human sweat and hard work—an energy flow—can put a disarrayed house back in order. Recognize, though, when a house is so reordered, it's done at the expense of the increasing disorder of the humans cleaning the house; after all, tidying a household is a tiresome activity (especially when a house is populated, as in my case, by curious and exploratory children), often making us feel listless for want of energy. In turn, we humans can become reinvigorated (i.e., personally energized or ordered) by eating food—also an energy flow—but this renewed order is, further in turn, secured at the expense of the solar energy that helped initially produce the food. I shall have more to say about energy consumption as a means of reordering systems in the last section of this chapter; for now, let me stress that this emergence of order from chaos is not a violation of the second law of thermodynamics.

Earth as a whole has a stock of natural energy resources. Most of the familiar ones—oil, gas, coal—are renewable over geological time scales of millions of years, but must realistically be treated as nonrenewable on the century time scales of humans and even the millennial time scales of civilizations. If natural resources were the only sources of energy available on Earth, and no sources existed outside our planet—a closed Earth system—then, when these sources were completely depleted, there would be no differences in energy states and Earth would eventually reach equilibrium; all parts of Earth would have attained the same temperature and Earth's entropy would then be maximized.

In point of fact, Earth itself is not a closed system, nor has our planet yet reached equilibrium; our world is still in a formative, albeit prolonged, stage of development. Additional matter and energy, especially sunlight, reach Earth's surface daily. Though essentially fixed in rate and pattern of arrival at Earth, solar energy is for all practical purposes unlimited. Long after terrestrial stocks are exhausted, our Sun will still act as a powerful source of energy, a cosmic hearth pouring forth heat and light, thus prohibiting Earth from reaching a state of equilibrium. Over the truly long haul, however, the Sun too will someday run out of fuel, as even today it degrades its own energy supply with each passing second. So, in five billion years or so, we can be reasonably sure that Earth will have attained equilibrium, its entropy maximized.

Without some other source of additional matter or energy, the

Sun itself will eventually equilibrate, the temperature of its matter everywhere the same and the entropy of the entire Solar System maximized. Though nuclear burning in the Sun is scheduled to cease several billion years hence, such a burned-out star will probably not reach a genuine equilibrium for many tens, even hundreds of billions of years thereafter. Truly great amounts of time are needed for a dead star to pass through the white-dwarf, red-dwarf, and brown-dwarf stages of stellar evolution, eventually becoming a black dwarf—a dark, decrepit clinker in space.

To be sure, the second law of thermodynamics has universal applicability. Not merely dictating the evolution of Earth and Sun, the second law also applies to stars and galaxies, indeed to the cosmos as a whole. The entire Universe currently races outward from the initial Big Bang explosion, cooling, thinning, and inevitably developing ever-smaller differences among the energy states of its multivaried contents. This eventual state of maximum entropy represents a "death" (some say a "heat death," but it's really a "cold death" by human physiological standards); at some time in the distant future all available energy in the Universe might become expended, thus rendering further activity impossible. All galaxies, stars, planets, and life forms will have decayed—even the most elementary atomic motions will have ceased—as the ultimate equilibrium is achieved. Clausius called it an "unchangeable death," a state of maximum disorder in which all organization and structure will have fully disintegrated and life itself will no longer be possible. As such, cosmic evolution would cease, for this is truly a state of eternal rest.

> *Time present and time past*
> *Are both perhaps present in our future*
> *And time future contained in time past*
>
> . . .
>
> *If all time is eternally present*
> *All time is unredeemable*
> —*T. S. Eliot*

The notion of time is basic to rational inquiry, now as well as in the past. Aristotle, among many Greeks of antiquity, regarded time as an absolutely fundamental concept. Mimicking Heraclitus, who taught that the Universe is the totality of *events*, not of *things*, Aristotle believed that temporal flux is an intrinsic feature of the ultimate basis of nature. And more than two millennia later Einstein warned that time should not be relegated to a poor second; time deserves at least equal footing with the concept of space, thus forming a spacetime continuum. As "geometry" deals with space, perhaps we need a subject of "chronometry" to deal with time.

But what makes time, an entity that cannot be touched, smelled, or transferred, as much a fiber of our Universe as hydrogen or helium? Is the issue of time so basic to our thinking that we are in danger of overlooking some major assumptions in our discussions of evolution? And is our perception of the direction of time arbitrary?

The concept of time is indeed essential to the subject of cosmic evolution and especially to the laws of thermodynamics that govern it. As one might expect, the most elementary notion of change cannot be understood without an analysis of time. In our Western, technologically oriented worldview, time "flows" in only one direction—forward. Alternatively stated: Time is irreversible; "time marches on." Expressed still another way: Time moves in whatever way entropy increases; entropy was less in the past and most assuredly will be greater in the future. Eddington put it succinctly: "Entropy is time's arrow."

(We must be careful here since the fact that galaxies, stars, planets, and life forms locally decrease entropy doesn't mean that time is reversed. As I shall subsequently stress in the latter parts of this chapter, the larger environments surrounding these objects still experience an increase in entropy. And the net entropy change for any system plus its environment is always positive.)

In a sense, then, the second law of thermodynamics points the direction of temporal change, although it gives no indication of the speed at which change occurs. Nor does it elucidate much about that moment we subjectively call "now" which moves inexorably into the future. According to biologist Harold Blum, the second law "is time's arrow, not time's measuring stick." Nor is the flow of events along time's arrow likely to be constant or uniform; some events—supernova explosions, airplane crashes, the swat of a fly, among others—proceed more rapidly than others in the direction of greater randomness. Thus, although increasing disorder can be generally taken as a measure of the *direction* of time, it cannot be taken as a measure of the *rate* of time's passage. The rate of change differs for each specific event.

The fact that I personally prefer to represent diagrammatically the salient features of cosmic evolution astride an actual arrow of time is not significant; nor is the linearity of the arrow or its left-to-right direction. The arrow merely provides a convenient intellectual road map of the *sequence* of major events in the history of the Universe. Others may prefer to visualize time moving abstractly from right to left, or up and down, or even having the shape of a knotted, open-ended pretzel. Provided the ordering of events accords with observation—in the main, galaxies originated first, followed by stars, planets, and life forms—then the arrow can be drawn with

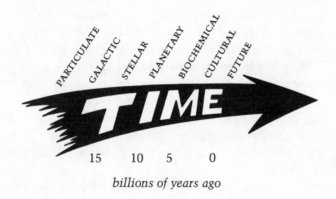

billions of years ago

. . . a convenient intellectual road map of the sequence *of major events in the history of the Universe.*

arbitrary shape and orientation. Furthermore, the arrow should be imagined to be flexible, permitting adjustments of the timing of historical events (as new knowledge accumulates) without upsetting the principal temporal sequence. (In this respect, it hardly matters if the Universe is as young as ten billion years or as old as twenty billion years; accordionlike modifications in the arrow of time can expand or contract history according to the latest data, while still preserving the successive order of cosmic organization.)

To many researchers, the arrow of time is a direct consequence of the expansion of the Universe. Given the contrast between hot stars and cold surrounding space, the Universe en masse is far removed from a state of thermodynamic equilibrium. Because of its very expansion, the dark realms of the Universe are perfect sinks—cold reservoirs into which stellar photons irreversibly flow.

Other researchers look to various evolutionary records for insight into the meaning of time's arrow. For example, fossils embedded in rocks imply that complex life arose from simple life. Moreover, such evolutionary records are produced not only by biological systems but also by inanimate objects. The stratigraphic ordering of Earth's rocks is widely regarded as a chronological record. The Moon's cratered surface provides a history of its past. The changing internal structures and chemical compositions of stars record the process of their aging. And the varied forms among the clustered galaxies suggest evolutionary events that granted them shape. Still, it is the expansion of the Universe that provides a temporal perspective for all inorganic evolution, which ultimately gave rise to all organic processes, including biological evolution, which in turn now culminates in our conscious awareness of time itself; hence, the reason I suggested in the Prologue that "cosmic evolution" is hardly more than a fancy term for universal change throughout eternity.

Furthermore, these observed evolutionary trends seem to be irreversible. Although fossil remains of both vertebrates and invertebrates imply that structures or functions once gained can be lost, structures that are lost can seldom be regained. Evolution is a consequence of many variations, chancy to be sure, but occurring in a definite order, and for it to be reversible, a highly improbable recurrence of specific variations would need to act in an inverse fashion to those that brought about the original transformations.

Complexity itself, however, is insufficient to demonstrate the direction of evolution. For not all species have become increasingly complex; sponges, roaches, spiders, bees, among myriad invertebrates, are trapped in an endless cycle of perfected day-to-day routines and have thus remained virtually unchanged for eons. Instead, I suggest that the "direction" of evolution obeys the following basic maxim: The collective efforts of living organisms tend to maximally utilize (per unit volume) both their energy intake from the

Sun and the use of free energy by dissipative (dispersive) processes occurring within them. It is in this sense that the biological evolutionary process has some temporal direction.

Likewise, the release of energy from the Sun and stars suggests a unilateral process. Solar heat and starlight are created and maintained by the conversion of gravitational into nuclear energy. Depending upon the mass of the star, the fusion reactions continue steadily for billions of years, but without any known compensating process capable of refueling cosmic objects, fusion cannot continue indefinitely. Stars too are not eternal. And their finite, albeit astronomical, lifetime is enough to impart a temporal history to the Universe. Indeed, the above-stated maxim applies to all of cosmic evolution: Organized structures everywhere seem to maximally utilize (per unit volume) the flux of energy passing through them.

On both terrestrial and celestial scales, then, we have abundant evidence of a temporal trend in the Universe—provided we consider sufficiently long intervals of time. Macroscopic events in nature seem everywhere to be irreversible.

So, while some physicists seek to prove that time's arrow can be derived from something more fundamental like entropy, others strive to describe it in terms of the flow of free energy through systems that are "open" to their surrounding environments. And clear across the spectrum of human scholarship, some thinkers, especially philosophers, deny that time is real; thus spoke Lucretius, representative of the ancient world: ". . . time itself does not exist; but from things themselves results a sense of what has already taken place, what is now going on, and what is to come." Whoever is correct, perhaps the essence of time is its transience, an ephemeral quality that will likely prevent its nature from ever being explained in terms of anything more fundamental. As G. J. Whitrow has put it, "Time is the mode of activity, and without activity there can be no time. Consequently, time does not exist independently of events, but is an aspect of the nature of the Universe and all that comprises it."

> *Enough if we adduce probabilities.*
> *—Plato*

> *Chance favors the prepared mind.*
> *—Louis Pasteur*

There are three kinds of lies—
lies, damn lies, and statistics.
                    —Mark Twain

Classical, deterministic physics has now fallen, and with it the mechanical Newtonian concept that every event can be precisely described. Whereas Newton's view stipulates the physical world as a closed system dominated by cause and effect, we now acknowledge the apparently insurmountable problem that the basic data for *every* particle of matter cannot be completely specified. After all, a mere drop of water contains some 10^{21} (a billion trillion) molecules.

Furthermore, early in the twentieth century, when scientists began probing the microworld, trying to locate, isolate, and measure the elementary particles of matter, they were surprised to learn that the realm of the atom is fundamentally different from the realm of terrestrial familiarity. Largely through the efforts of the German scientist Werner Heisenberg, the pioneering quantum physicists of the mid-1920s reached a consensus that objective observations of nature's most basic entities are impossible. The very act of atomic and subatomic observation interferes with the process of measurement, significantly altering the state of the object observed. In our attempt to decipher natural order, we have become, as the great Danish atomic physicist Niels Bohr put it, "both spectators and actors in the great drama of existence." We simply cannot separate ourselves from the world around us, regardless of how hard we try.

Whereas classical physics sought to determine precise values for certain pairs of physical variables—for instance, the position and momentum of a particle—quantum physics postulates that these properties can be simultaneously observed only to within a certain limiting accuracy. One or the other can be measured arbitrarily well at any moment—but not both. To perceive the quantum microworld, we must scatter other subatomic particles from those being studied. In doing so, physicists have invariably found that the measurements fluctuate about an average value and that these fluctuations arise not so much from practical imperfections of the experimental equipment as from the fuzziness introduced by the act of measurement itself. The resulting premise—the Heisenberg Uncertainty Principle—to this day inherently limits the goals of classical physics. Gone is the deterministic and mechanistic world paradigm, as well as the notions of strict causality and predestination, that characterized physics for several centuries since Bacon, Descartes, and Newton. Gone, also, are the idea of the future as implicit in the past and the notion that there is no essential novelty in the world. No longer can we predict, given an initial set of conditions, the one and only final state of some event. In quantum physics there is no single final state but only several possible alternative states. Probabilities can be assigned to each of the possible outcomes if we are given an initial set of conditions, but the outcome is not fixed or predetermined. Even Einstein was apparently wrong; God *does* seem to play dice with the Universe, thereby creating much originality.

Of course, averages can be established within the context of large numbers. In this way we can make statistical statements with a high probability of success. For example, we can derive statistical averages for a stable (equilibrium) system that suffice to predict that system's macroscopic behavior with virtual certainty. By contrast, as we shall see toward the end of the chapter, this is not necessarily true for unstable systems. Such a nonequilibrium system is affected by microscopic disturbances that can accumulate so that events on the macroscopic level reflect chance activity at the microscopic level. Expressed another way: While the Uncertainty Principle was once thought to be of no importance for the description of macroscopic objects, such as living systems, recent studies of the role of minute fluctuations in nonequilibrium systems imply that this is not the case; randomness seemingly retains some influence on the macroscopic level as well. Personally I'm fascinated by the idea that the chance behavior of individual particles in the microdomain can yield observable macroscopic changes, but more on this later.

Chance. Probability. Statistics. Doubtless chance is a factor in all aspects of cosmic evolution, but it cannot be the sole instrument of change. Take, for instance, the issue of galaxy formation. Even Newton knew that a huge cloud of gas will, naturally and of its own accord, randomly develop density inhomogeneities here and there. (". . . if matter was evenly disposed throughout an infinite space . . . some of it would convene into one mass and some into another, so as to make an infinite number of greater masses scattered at great distances from one another throughout all that infinite space.") By chance, and chance alone, the cloud's matter spawns clumps that become the seeds for the galaxies themselves. However, if we expect the total number of atoms ($\sim 10^{68}$) required to fashion a typical galaxy to be collected exclusively by random encounters of gas particles, then we run into a problem. A chance accumulation of this vast quantity of atoms takes several tens of billions of years; in view of the Universe's current age of some fifteen billion years, there should therefore now be no such galaxies. The observational evidence that galaxies do in fact exist, and in no small numbers, strongly implies that chance cannot be the only factor governing the origin of these grand systems. Chance plays a role, to be sure—especially in the initial process that triggers the fragmentation of the cloud—but other agents, such as turbulence and shocks, must accelerate the growth of the inhomogeneities so that myriad galaxies can form within a time scale shorter than the age of the Universe; in fact, the enhancement process must be surprisingly efficient since observations further demonstrate, as noted in Chapter 1, that all galaxies are old.

Formation of the precursor molecules of life's origin provides another opportunity to illustrate the limited role of chance in nature. As also described in Chapter 1, simple molecules such as ammonia, methane, water vapor, and carbon dioxide interact with one another in the presence of energy to generate larger molecules. The end products are not just a random assortment of molecules; they comprise the two dozen amino acids and nucleotide bases common to all life on Earth. And regardless of how this chemical evolutionary experiment is performed (provided the gases simulating our primordial planet are irradiated with realistic amounts of energy in the absence of free oxygen), the soupy organic matter trapped in the test tube always yields the same relative proportions of proteinoid compounds. Though it makes an awful mess, I've even done this experiment in my bathtub with essentially the prescribed results, as has Julia Child in her kitchen while concocting her "primordial soup" recipe. The point is that if the original reactants were re-forming into larger molecules by chance alone, the products would be among billions upon billions of possibilities and would likely vary each time the experiment was run. But the results of this experiment show no such diversity. Of the myriads of fundamental organic groupings and compounds that could possibly result from the random combinations of all sorts of simple atoms and molecules, only about fifteen hundred are actually employed on Earth; and these groups, which comprise the essence of terrestrial biology, are in turn based upon only about fifty simple organic molecules, the most important of which are the above-mentioned acids and bases. Some factor other than chance is necessarily involved in the prebiotic chemistry of life's origin, though one need not resort to supernatural phenomena. That other factor we reason to be the microscopic electric forces naturally at work among the molecules—forces that guide and bond small molecules into the larger clusters appropriate to life as we know it, thus granting the products some specificity and stability. A benzene ring, for instance, is a good deal more stable than a linear array of the same atoms and molecules. And it doesn't take long for reasonably complex molecules to form, not nearly as long as probability theory predicts by a chancy assembly of atoms. In short, the well-known electromagnetic force acts as a molecular sieve or probability selector, fostering only certain combinations and thereby guiding organization from amidst some of the randomness.

Molecules even more complex than life's simple acids and bases also cannot be synthesized by chance acting alone. For example, the simplest protein, insulin, comprises fifty-one amino acids linked in a specific order along a molecular chain. Using probability theory, we can estimate the chances of randomly assembling the correct number and order of acids; given that twenty amino acids are involved, the answer is $1/20^{51}$, which equals $1/10^{66}$. This means that the twenty acids must be randomly assembled 10^{66}, or a million trillion trillion trillion trillion trillion, times before we get insulin. As this is obviously a great many permutations, we could randomly assemble the twenty amino acids trillions upon trillions of times per second for the entire history of the Universe and still not achieve *by chance* the correct composition of this protein. And clearly, to assemble larger proteins and nucleic acids, let alone a human being, would be vastly less probable if it had to be done randomly, starting only with atoms or simple molecules. Not at all an argument favoring creationism; rather, it is once again the natural forces of order that tend to tame chance.

In regard to life itself, the process of natural selection further acts as a sieve or sifting mechanism, permitting some species to thrive quite naturally even while others normally perish. Chance admittedly provides the raw material for biological evolution, but natural selection is a decidedly deterministic action that directs evolutionary change. Indeed, contrary to popular opinion, Darwin never said that the order so prevalent in our living world arises from randomness per se. Yet even the strictly limited role of chance in modern Neo-Darwinism, when coupled with the principle of natural selection, is capable of generating highly improbable results. Theodosius Dobzhansky, one of the twentieth century's foremost biologists, has addressed this delicate interplay of chance and necessity, of mutation and selection: "Evolution is a synthesis of determinism and chance, and this synthesis makes it a creative process. Any creative process involves, however, a risk of failure, which in biological evolution means extinction. On the other hand, creativity makes possible striking successes and discoveries."

We are left with the notion that only when vast numbers are involved do events begin to show predictable patterns, thereby comprising what we recognize as the (terrestrially familiar) behavior of matter in the macrocosm. In this way, we can understand how random processes can yield highly predictable order—though only within the frame of large numbers and repeated trials—much as probability theory can elaborately predict the results of numerous coin tosses, the archetypical random event. Even so, a price is paid: Whenever a system is represented by an average, we inevitably lose some information about the total system. Consequently, we need to specify the number of individual cases used to determine a particular average; this is a measure of the information lost when working with averages.

These statements can be clarified rather simply by use of the modern subject of *statistical* mechanics—a probabilistic interpretation of classical mechanics—which supplements the materialistic physics of old. Invented about a century ago by the Austrian Ludwig Boltzmann and the first great American theoretical physicist J. Williard Gibbs, statistical mechanics (really statistical thermodynamics) employs large aggregates of particles to represent thermodynamic concepts. Of paramount import, in 1872 Boltzmann announced his famous formula (subsequently engraved on his tombstone in Vienna), S = k lnW. Here S denotes entropy, k is a proportionality factor now known as Boltzmann's constant, ln is a mathematical symbol for the natural logarithm, and W is the number of different arrangements of the individual parts or "microscopic states"—positions, velocities, and various quantum properties—of a given macroscopic system. Quantitatively this basic formula states that the entropy is determined by the number of possible states (or, more accurately, by the logarithm of this number). To appreciate how the formula works, consider the following hypothetical case.

Imagine two bodies (1 and 2) in contact. Assume further that they both are immersed in a perfectly insulating container, so they can exchange heat—but only heat, no matter. Now, at any instant, each body will have some energy, say E_1 and E_2. The first law of thermodynamics then demands that the sum of these two energies, $E_1 + E_2$, is constant at all times. Still, at any one time, there will be W_1 states of the microscopic entities comprising Body No. 1 and likewise W_2 microscopic states for Body No. 2; here W_1 is a function of E_1, and W_2 a function of E_2. The total number of states for the *whole* system is then W = $W_1 \times W_2$, a multiplicative result since *each* state in Body No. 1 can be associated with *each* state in Body No. 2. (The multiplicative nature of W, in contrast with the additive nature of E [or S], accounts for the logarithmic function in the above Boltzmann formula. It's not unlike the way we evaluate betting odds at the racetrack; we know that our chance [probability] of picking a given horse to "win" *or* "place" is the sum of the individual odds, whereas our chance of picking two horses to "win" *and* "place" is their product.)

The role played by statistics can be further appreciated by recognizing that the number, W, of microscopic states essentially measures the probability, P, of the occurrence of those states. This is true because the most probable configuration of the particles in a gas is the high-entropy state for which they are evenly mixed up or "disordered"; likewise, during a poker game, the likelihood of drawing a desirable hand is small since the entropy of such a preferred or "ordered" set of cards is considered low. Continuing our discussion of the above two-body system, we then reason that the most probable energy distribution is that for which W is maximized; this occurs when the temperatures of the two bodies are equal. Thus, we can rewrite the second law of thermodynamics as $S = k \ln P$ and interpret it in the following way: Any closed, or isolated, system automatically tends toward an equilibrium state of maximum probability—namely, an equilization of temperature, pressure, and so on. And since the probabilities of having ordered molecular states (e.g., where molecules in one part of the system have one temperature while those in the remaining part have another) are far less than those of random or disordered states, Boltzmann's law then signifies that ordered states tend to degenerate into disordered ones. This law also explains why any material system's available energy tends to diminish with time; useful energy is orderly energy, whereas heat, associated with the random motions of large numbers of molecules, is disorderly energy. Therefore, the energy of any closed system indeed remains constant but tends to become progressively less free to perform useful work as more of it changes into heat, thereby increasing the disorderliness of its component parts.

This is how the principle of increasing entropy has come to be regarded as a measure of the disorder of a system. The notion of increasing entropy has ceased to be an invariable law of nature; rather, it is a statistical law. Accordingly, a reversal of the usual trend—toward a state of lower entropy—is no longer impossible but only highly improbable. For example, a portion of water in a kettle atop a fire could (theoretically) freeze while the remainder boils, although this is so unlikely as to be considered impossible in our practical experience.

Thus, the static Newtonian idea that treats all phenomena as stable, fixed, isolated components of matter has given way to the notion that all things are intrinsically unstable and share part of a dynamic flow. Much like Heraclitus, we no longer regard things as "fixed" or "being" or think that they even "exist." Instead, everything in the Universe is always in the act of "becoming." All things—living and nonliving alike—are continually changing.

Now, since all things are made of matter and energy, both of which constantly change, we can conclude that the process of becoming is governed by the laws of thermodynamics. As noted earlier, the second law determines the *direction* in which matter and energy change (or flow), but it cannot determine the rate of that change. In nature this rate fluctuates, within recent Earthly times

. . . increasing entropy has ceased to be an invariable law of nature; rather, it is a statistical law.

largely because of the cultural impact of human beings. To be sure, there is nothing smooth about the ebb and flow of the becoming process. Change proceeds in jumps and spurts or, to use a fashionable term borrowed from recent studies of biological evolution, in a "punctuated" manner. Accordingly, we say that the process of change is characterized by nonequilibrium thermodynamics, and we use this relatively new discipline to predict scenarios of what *might* happen during any given event, what *might* result from any given change.

> *Order. We must have order—*
> *but not too much order!*
> —*A. N. Whitehead*

As noted in the previous section, the thermodynamicist describes order in terms of the arrangements of the numerous microscopic states comprising the various macroscopic properties of any system. The larger the number of possible arrangements, the greater the disorder or randomness. The smaller the number, the greater the order, for we can think of order as a form of restraint on the way in which a system can arrange itself. As an example, when a system has only a single possible arrangement for its microscopic states, Boltzmann's entropy formula reduces to $S = k \, ln1 = 0$, meaning that the system has zero entropy—admittedly an idealistic case since only, for instance, the severely impaired behavior of the atoms and molecules in a perfectly ordered crystal at the unattainably lowest possible temperature (i.e., 0 degrees Kelvin or -273 degrees Celsius) could truly have zero entropy.

The statistical character of the second law arises from the fact, as also noted earlier, that the laws of probability govern the direction in which natural processes occur. This is perhaps best understood once we realize that the microscopic properties of individual molecules cannot be directed or controlled by some human agency. The classical example of the "Maxwell demon" is an appropriate case in point. As first suggested by the nineteenth-century Scottish physicist James Clerk Maxwell, suppose that a vessel containing a gas were divided into two compartments by a partition having a trapdoor controlled by a superhuman demon or robot that can distinguish individual molecules. The proposition is that such a microscopic robot could open the trapdoor only when fast-moving molecules approached from one side of the door, thereby allowing the fast molecules eventually to collect in one compartment and the slow ones in the other. Without having expended any work or applied any energy, this exercise would thus raise the temperature of the compartment housing the swifter molecules while lowering the other—namely, cause a change from disorder to order, in apparent violation of the second law.

Doubtless we could squirm out of Maxwell's paradox by affirming the practical impossibility of such demonic robots. But such an assertion would merely dodge the issue, admitting its possibility in principle, even if not in practice. The paradox is solved upon the realization that even such a hypothetical robot would need to know when to open and close the trapdoor, for knowledge of the velocity of each approaching molecule is required if a segregation is to result ultimately between the fast- and slow-moving molecules. In short, the sorting demon would need to have a way to obtain *some information*. For instance, we might imagine that a microscopic flashlight (or radar beacon) could be used to measure each molecule's velocity, thus providing the robot with the essential data about whether or not to open the door. This method would work at least in principle, but it introduces external energy into the vessel. After all, the energy needed to operate the flashlight (or any other apparatus capable of providing the needed information) must come from outside the enclosed vessel. Alas, even a hypothetical robot *in a closed system* could not arrange for a separation of the molecular velocities. Thus, once again we conclude that an energy flow through an open system is an absolute necessity if order is to be created from disorder. And provided we view the process of change sufficiently broadly, then there is no contradiction of the fundamental laws of thermodynamics.

Our detour into the world of Maxwell's demon was not for naught, for the "information" needed to resolve the paradox points the way to a reconciliation of an apparent conflict between the evident "destructiveness" of the second law of thermodynamics and the observed "constructiveness" of cosmic evolution.

In recent years the concept of entropy has been used to illuminate some aspects of the subject known as "information theory." As generally noted earlier in this chapter, entropy can be taken to be a measure of the lack of information about the internal structure of a system. This lack of information admits a great variety of possible structural arrangements among the system's microscopic states; the positions, motions, energies, etc. of the many varied parts of a high-entropy system cannot in practice be specified clearly. Since any one of these microscopic states might be realized at any given time, the lack of information (or ignorance) about a system corresponds to what we have earlier labeled "disorder" (or "uncertainty") in the system. For example, when a system is at equilibrium—the state of greatest change on the microscopic scale but of the greatest uniformity as seen by the human observer on the macroscopic scale—we have the least possible knowledge of how the various parts of the system are arranged and where each one is and what it is doing. At equilibrium, all such microstates have equal (large) probability, P; that is why, in information theory, the information itself, I, varies inversely as the probability: specifically, $I \propto ln\ (1/P)$, which resembles the equation noted for entropy in the previous section of this chapter.

Of course, the word "information" can connote different ideas in different contexts. In ordinary speech we use this word as a synonym for news, knowledge, intelligence, and so on. But in the more purified world of the communications engineer (or should I say the rarefied air of the cyberneticist), stress is placed on the quantitative aspect of the *flow* of an intangible attribute (called "information") from transmitter to receiver. It is in this latter context—a message—that we can sense a connection between the sciences of thermodynamics and information theory. To see this, consider how the concept of entropy has been applied to the transmission of information by electronic means, especially in connection with the use of telephones, radios, computers, and the like. As is familiar to all of us, such transmissions can be adversely affected on occasion by errors and interference of various kinds, the result often being that the final message is less accurate or conveys less information than the original message. This is true whether the information is transmitted via noisy telephones, staticky radio sets, distorted stereo systems, faulty telegraph lines, snowy television screens, poorly networked computers, or a host of other means to move information from one place to another. Communicating a message inevitably results in a change to a state of greater inaccuracy; the integrity of the message is unavoidably corrupted, some information invariably lost. The same applies to the printed media; when newspapers are torn, letters soiled, or books burned, the informational content of the original message declines.

Now if we associate the loss of information with a decrease of order, or likewise an increase of disorder, and if we recall from immediately above that the equation for information equals the inverse of that for entropy, then we surmise that a gain of information is directly related to *negative* entropy. This last pair of words plays such a pivotal role in information theory that they are often described by a single peculiar term, "negentropy." Expressed another way: When a system is orderly (that is, low in entropy and rich in structure), more can be known about that system than when it is disorderly and high in entropy. And if entropy measures disorder or the lack of information about a system, then negentropy must be a valid measure of the order or presence of information; each of these latter terms—order, negentropy, information—is essentially synonymous. In short, if we gain some information about a system, the previously existing uncertainty about that system is diminished. Said G. N. Lewis, one of the foremost American chemists of the early twentieth century: "Gain in entropy always means loss of information, and nothing more. It is a subjective concept, but we can express it in its least subjective form. . . ."

Thus, we reason that the external energy needed to operate the optical or radar apparatus used to secure the information for Maxwell's demon effectively poured negentropy into the vessel, thereby creating order by segregating the swifter gas molecules from their more sluggard companions. We shall return in the latter part of Chapter 4 (and in the Appendix) to make use of this strong correlation between the concepts of entropy change in thermodynamics and of information exchange in transmission theory.

> *Beyond that which is known and what is thought real,*
> *lies a realm just as real but more complex.*
> *—attributed to J. B. S. Haldane*

A natural basis for the creation and maintenance of ordered structures in the Universe has occupied the minds of humans since the beginnings of recorded time; this much is reasonably clear from the historical outline of Chapter 2. Life forms, in particular, are highly complex and intricate systems whose substantial order implies great organization—structure gradually acquired by means of long periods of evolution. Yet despite the obvious organization characterizing life in general, there must be a regular procession of energy and matter into and out of living systems. Otherwise, life's order could be neither created nor maintained. Similar statements regarding energy consumption and information growth hold equally valid for the ordered structures that are galaxies, stars, and planets.

That this must be true follows from our previously noted (and tested) idea that the usual course of thermodynamics tends to lead systems toward an equilibrium state of maximum disorder. For any isolated system unable to exchange energy and matter with its surroundings, this tendency is expressed in terms of the now-familiar entropy, which, as noted earlier in this chapter, increases until maximized. In short, thermodynamics' second law effectively rules out ordered structures in isolated (closed) systems. Consequently, we conclude that the apparent contradiction between the observed universal order and the theoretical laws of physics cannot be resolved in terms of the usual methods of equilibrium thermodynamics or even equilibrium statistical mechanics.

To associate order with the formation of low-entropy structures such as macromolecules, cells, and even whole planets, stars, and galaxies, we need to consider the physics of nonequilibrium open systems. From simple bacteria to complex humans, from round stars to spiral galaxies, ordered systems must be maintained (and in some cases reproduced) by means of a continuous exchange of energy and matter with their surrounding environments. It is because of this inward and outward flow or flux of matter and energy that a system is considered "open."

Championed in recent years by the Belgian physical chemist Ilya Prigogine, several (largely European) groups are now trying to model open systems. Called "dissipative structures," all ordered objects—living and nonliving—apparently maintain their beings by means of a continual flow of available energy from the outsides to their insides. The more complex and intricate the structure, the more energy intake (per unit mass) is needed to maintain itself. In the process these structures can transfer some of their entropy (or

136

dissipate some of their energy; hence, their name) into the external environment with which they interact. Accordingly, order is maintained by a steady consumption of substances rich in energy, followed by a discharge of substances low in energy. (Either "substance" could be pure energy [i.e., radiation] or matter itself.)

Now we know well that fluctuations—random deviations from some average, equilibrium value of density, temperature, pressure, etc. (also called "instabilities" or "inhomogeneities")—are a common phenomenon; they inevitably occur in any system having many degrees of freedom. Normally, as in equilibrium thermodynamics, instabilities regress in time and disappear; even in a closed system, such internal fluctuations can generate local, microscopic reductions in entropy, but the second law ensures that they will always balance themselves out. Nor can an open system *near equilibrium* evolve spontaneously to new and interesting structures. But should those fluctuations become too great for the open system to damp, the system will then depart far from equilibrium and be forced to reorganize. Prigogine and his colleagues maintain that such reorganization tends toward a higher order of complexity, provided the amplified fluctuations are driven and stabilized by the flow of energy and matter from the surroundings. Furthermore, since each successive reordering causes more complexity than the preceding one, such systems become even more susceptible to fluctuations. Complexity itself consequently creates the condition for greater instability, which in turn provides an opportunity for a greater reordering. The resulting process—termed "order through fluctuations"—is a distinctly evolutionary one, complete with feedback loops that drive the system farther from equilibrium. And as the energy consumption and resulting complexity accelerate, so does the evolutionary process.

The important point is that beyond some instability threshold, physical systems can foster the spontaneous creation of an entire hierarchy of new structures displaying surprising amounts of coherent behavior. Such highly ordered, dissipative structures can be maintained only through a sufficient exchange of energy and matter with their surroundings; only this incoming flux can support the high degree of organization needed for existence. The work needed to maintain the system far from equilibrium is the source of the order. An apt analogy often made is that of urban existence: A city can survive only as long as food, fuel, and other vital commodities flow in while products and wastes flow out. Even Freud apparently regarded mental urges and instincts (ranging from ordinary thoughts to exotic dreams) as gusts of energy swirling through the brain.

For open systems, then, we must reformulate the second law of thermodynamics. Specifically we now recognize two kinds of entropy: one, the normal (positive) entropy production inside the system caused by irreversible events, such as friction, chemical reactions, conduction of heat, etc.; the other, entropy production caused by exchanges of the system with the outside world beyond the system. For a closed system this latter entropy contribution is zero. But for an open system the entropic change caused by environmental interaction can be either positive or negative. The sign of the contribution (plus or minus) is generally undetermined and depends entirely on the nature of the energy flow across an open system's boundary. Accordingly, in principle at least, an open system can establish and sustain a negative entropy (or negentropy) change vis-à-vis its environment, and during the course of evolution may reach a state wherein its entropy has physically decreased to a smaller value than at the start. (Note that I am not stating that the rate of entropy growth becomes lessened or even reduced to zero; under appropriate conditions, the entropy itself within open systems can be actually reduced.) In this nonequilibrium fashion, order can be achieved within a system by a sort of spontaneous self-organization.

Furthermore, such open systems can be prevented from reaching equilibrium by the regular introduction of fresh reactants and by the regular removal of the derived products. In point of fact, such a system can be regulated to achieve a constant ratio of reactants to products. The system then *appears* to be in equilibrium because this ratio does not change over time. But in contrast with true equilibrium (where entropy is maximized), this kind of process continually produces entropy and dissipates it into its surrounding environment.

The phenomenon of refrigeration provides an especially simple illustration of the creation and maintenance of nonequilibrium order within an open system at the expense of increasing disorder outside that system. Imagine a container of water saturated with sugar. Normally such a solution comprises a relatively random state, the water and sugar molecules freely able to move about, thus

. . . an open system can establish and sustain a negative entropy (or negentropy) change vis-à-vis its environment . . .

occupying a great many positions relative to one another. If we allow this system to cool, the sugar crystals begin to form spontaneously; they become highly organized, with the individual molecules occupying rather exact positions in the emerging matrix comprising the crystal. Having order, the newly formed crystals themselves necessarily possess lower entropy than the surrounding solution— that is, the act of crystal formation decreases the entropy in certain localized parts of the system. We should then expect to find an increase in entropy somewhere else. The solution itself is unlikely to have an appreciably increased entropy, for its temperature has also been lowered. Thus, to find where the entropy has increased, we need to enlarge the system beyond the confines of the vessel containing the solution. Indeed, as the solution cools, heat flows to the surroundings beyond the container. Therefore, it is the air outside the container that must suffer an increase in entropy—which is exactly why an open refrigerator cannot be used to "air-condition" a kitchen in midsummer; a refrigerator, while cooling its contents, in fact tends to warm the kitchen.

Thus, while the *destruction* of order always prevails in a system in *or near* thermodynamic equilibrium, the *construction* of order may occur in a system far from equilibrium. Whereas traditional thermodynamics deals with the first type of physical behavior, new techniques only now being developed are needed to decipher the second type of behavior.

Heating of a fluid from below provides a good illustration of such dual behavior, a case technically termed the "Bénard hydrodynamics problem" (and named for the French physicist who studied this experiment extensively in 1900). Such externally applied energy generates a vertical thermal gradient in the fluid capable of amplifying any random molecular fluctuations. When the heating is slight (i.e., below some critical temperature or instability threshold), the energy of the system is merely distributed (by conduction) among the random thermal motions of the fluid's molecules and the fluid continues to appear homogeneous; the natural, random fluctuations are successfully damped (by viscosity), and the state of the system remains stable or equilibrated. But beyond this threshold (i.e., when the fluid is heated extensively), instabilities naturally become strengthened as large thermal gradients develop, thus spontaneously breaking the initial symmetry (or homogeneity) of the system; this causes the emergence of macroscopic inhomogeneities in the fluid—namely, small but distinctly and coherently organized convective eddies that can be seen with the naked eye. Anyone can verify this well-known phenomenon of convection by heating a pot of water on the stove; as the water is brought to near boiling, cells housing millions of H_2O molecules become buoyant, move systematically, and form upwelling patterns of a characteristic size. In this way, order can naturally emerge (that is, spontaneously self-organize) when the system is driven far beyond its equilibrium state. This is what is meant by the principle of "order through fluctuations" or "energy flow ordering."

Likewise, you can generate whirling eddies by slowly passing your hand through a tub of water. Rapid passage only produces splashing (a kind of turbulent chaos), but a steady movement yields organized whirlpools (or vortices) of swirling water in its wake. Here the tub of water may be considered an open system, with your hand providing some energy from outside. Without this energy the water would remain idle and quiescent, the epitome of a closed system in perfect equilibrium. But the application of external energy enables the liquid system to depart from equilibrium, to enhance fluctuations, and, so long as energy is provided, to create and maintain somewhat ordered structures.

Earth's weather is another practical example of a semblance of order within a complex and chaotic system of gas. Here our atmosphere is heated largely from the Sun-warmed crustal surface, thus creating thermal updrafts, surface winds, and other global meteorological phenomena in a reasonably systematic way. An especially apt case is the occasional yet sudden interruption of surface winds blowing across our planet's oceans. In the Canary Islands, for instance, the prevailing westerlies flowing across the Atlantic Ocean interact with the mountainous regions near Tenerife, causing swirling and turbulent eddies of moist air to form high in the atmosphere. (Such kilometer-sized vortices can be seen clearly by looking downward from orbit, though not easily by looking upward from the ground.) Solar-driven winds provide the external energy which, along with the fluctuation in the flow caused by the Canaries, eventually results in a growth of atmospheric structure. Very infrequently, should they be driven far enough from equilibrium and fueled with adequate moisture, such eddies can mature into full-scale hurricanes hundreds of kilometers across. Interestingly enough, the pancake shape, the spiral-arm structure, the disposition of energy, the differential rotation pattern, among many other morphological characteristics of hurricanes, bear an uncanny resemblance to those of spiral galaxies. Is it possible that the origins of nature's grandest structures were triggered in the early Universe as the rapid flow of radiation, launched by the Big Bang, swept past primordial gas fluctuations that acted as the turbulent seeds of galaxies? And since most meteorologists agree that some sort of turbulent "priming" is needed to initiate a hurricane, might not studies of the formative stages of such storms—in the very air we breathe—conceivably be used by astronomers to derive some clues to the elusive density inhomogeneities that presumably gave rise to protogalaxies more than ten billion years ago?

The laser provides yet another case of order's emerging when a system is driven far from equilibrium. "Laser" is an acronym for "light amplification by stimulated emission of radiation" and is a device used to produce high-quality (coherent) light. While an undergraduate student, I once built such an instrument out of, for the most part, a bed pipe and a discarded neon-sign generator. By energizing the atoms of carbon dioxide gas mixed with small amounts of nitrogen bled from the air into the long pipe, I was able to generate a plasma (i.e., ionized gas, the fourth state of matter). The resulting light depended critically upon the manner in which the incoming energy excited the gas. Below some instability threshold (a small application of energy as from an ordinary AC outlet), only incoherent light resulted, much like that from the ordinary light bulbs in our homes. But beyond that threshold (corresponding to a gradual increase above some critical value of the applied energy, then supplied by a more powerful generator), the plasma within the laser switched spontaneously to the emission of coherent light. The gaseous system is said to have changed from chaos to order, in effect punctuating its equilibrium to achieve a new level of organization.

Regarding each of these examples, Prigogine notes:

> The atoms and molecules in such systems must interact with their immediate neighbors through well-known, short-range chemical and physical forces. Each atom or molecule knows only its immediate neighbors and its direct environment. That

. . . near and far from equilibrium.

much is normal. But in these new [dissipative] structures, the interesting thing is that the atoms and molecules also exhibit a coherent behavior that goes beyond the requirements of their local situations . . . which increases in complexity and grows to be something vastly different from the mere sum of its parts.

We now regard nature's ordered systems as evolving through a series of instabilities. In the neighborhood of a stable (equilibrium) regime, evolution is sluggish or nonexistent since small fluctuations are continually damped; destruction of structure is the typical behavior wherein chaos rules. But near a transition threshold, evolution speeds up, and the final state depends on the probability of creating a fluctuation of a given type. Once this probability becomes appreciable, the system will eventually reach a unique steady state, in which construction of structure wherein order rules is distinctly possible. Such states can thereafter be starting points for further evolution characterized by perhaps even greater order and complexity.

Accordingly, we need not regard our observations of the evolution of order in the cosmos as conflicting with the central tenet of modern thermodynamics that stipulates chaos to be universally increasing. In particular, two different sets of physical laws are unnecessary to account for such differences in natural behavior. To be sure, most researchers have now reached a scientific consensus that the evolution of order is governed by only one physical principle, though one operating in demonstrably different physical situations: near and far from equilibrium. That single, unifying principle encompassing all elements of natural change is, once again, the second law of thermodynamics.

CHAPTER 4

Two Preeminent Changes

AS SENTIENT BEINGS WE HUMANS ON EARTH
NOW RESIDE AT THE DAWN OF A GREAT TRANS-
FORMATION IN THE HISTORY OF THE UNIVERSE

SEVERAL YEARS AGO, during a summer symposium that I was
colecturing to Harvard alumni, the Nobel physicist Steven Weinberg
argued that everything of importance had happened within the first
few minutes of the Universe. Claiming that nothing of appreciable
consequence had occurred thereafter, he reasoned that all subsequent events can be regarded as mere detail.

Many scientists, not least biologists I suppose, would regard
those views as somewhat provincial. To be sure, Weinberg was being
partly facetious, as his aim was to emphasize events of the early
Universe. For while relatively simple subatomic matter might well
have been created in the initial moments of the Universe, the more
complex and organized matter now surrounding us surely formed
well after its start. As I emphasized in my lectures to the alumni,

145

every dating technique developed by post-Renaissance science implies that complexity gradually arose from simplicity, clumpiness from homogeneity, or, as has been previously suggested, order from chaos.

Granted, an event of insuperable significance must have occurred when matter began to coagulate within the otherwise chaotic state of primal energy extant shortly after the Universe originated. This emergence of matter as the dominant constituent is indeed a preeminent change in the history of the Universe; I once regarded it as *the* preeminent event of all time. But as I noted earlier, we can now identify another fundamental transformation: one that occurs when technologically competent, intelligent life arises, in turn, from that matter. Our civilization on planet Earth is now slowly beginning to experience this second great change; though we hardly realize it, we now reside at the dawn of a whole new era—an era of opportunity that we ourselves have helped create.

> *Currents through time*
> *Offer threads to tie change,*
> *Spinning thoughts combine*
> *Matters of wide range.*
> *In cores of stars*
> *Coins are minted.*
>
> *Each breath of ours*
> *Volcanoes emitted.*
> *Overstuffed stars that expire*
> *Launch waves of shock,*
> *Undulations they inspire*
> *Terraform dust to rock.*
> *Investing time in the universe,*
> *Order comes out of chaos.*
> *Now always built from then.*
> *—Lola Judith Chaisson*

Contemporary cosmology is guided by two key observations: one of material objects, the other of radiative energy. The first concerns the recession of the galaxies, and the second a cosmic radio signal inundating all space. Although we now know that energy ruled the Universe prior to matter's current dominance, I judge it

146

pedagogically more meaningful to discuss initially the role of the material galaxies.

For much of this century, astronomers have realized that the galaxies (especially the cosmic hierarchy's even larger clumps of matter, the galaxy clusters) have some definite organized motion in space. And, I might add, in time, for the finite velocity of light guarantees that peering deeply into space is equivalent to probing back into time. The farther we observe material objects in space, the greater we penetrate the past. (Hence, my earlier claim that astronomers are the true savants of ancient history.)

Virtually all the galaxies are steadily receding, each one racing away from us with a velocity proportional to its distance from Earth. (I say "virtually" because a few *nearby* galaxies, including neighboring Andromeda, are known to have a component of their velocity toward us, but that's due to the random, small-scale motions that all galaxies display in addition to their more directed, large-scale recessional motions.) These two properties—velocity and distance—are highly correlated: The greater the distance of an object from us, the faster that object recedes, much like the expelled fragments of a detonated bomb. Presumably, the fastest-moving galaxies are by now farther away *because* of their high velocities. Visualizing the past by mentally reversing the outward flow of galaxies, we reason that all such galaxies were once members of a smaller, more compact, and hotter Universe. Accordingly, we surmise that an explosion of cosmic proportions—popularly termed the "Big Bang"— probably occurred at some time in the remote past. On the basis of the observed rate of the galaxies' recessional motions, we calculate that that time must have been about fifteen billion years ago (cf. the Appendix for a derivation of this number and for a discussion of its uncertainty). The galaxies, including our own Milky Way, share in the expansive aftermath of this cosmic bomb, for they delineate, at one and the same time, both the underlying fabric of the Universe and the scattered debris of that primeval explosion.

Recessional motions of the galaxies comprise our best evidence that the entire Universe is active. Like everything within it, the Universe changes with time. To be sure, the cosmos is expanding in a directed fashion; in short, it's evolving.

The second key observation pertinent to our modern cosmology concerns a radio signal that seems omnipresent in the Uni-

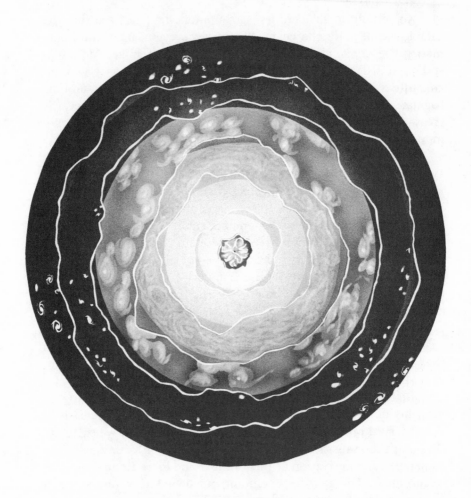

Visualizing the past by mentally reversing the outward flow of galaxies, we reason that all such galaxies were once members of a smaller, more compact, and hotter Universe. (In this diagram the three dimensions of space have been compressed onto the two dimensions of a sphere's surface, thereby allowing the fourth dimension—time—to be visualized radially progressing through a concentric series of such spheres.)

verse. Discovered some two decades ago serendipitously, this weak radiation floods every nook and cranny of space and sounds, to a radio astronomer, not much different from the hiss or static on an ordinary home (AM) radio receiver. (When converted into a video signal capable of display on a computer terminal, this ubiquitous radiation resembles the "snow" seen on a television tuned to an inactive channel.) Regardless of the direction observed or of the time of day, night, and year, this minute radio signal is ever-present, strongly suggesting its universal nature.

Static or not, this so-called cosmic background radiation contains much information whose subtleties we are still learning to extract. While the observed signal clearly arises from microscopic thermal events (i.e., elementary particles releasing energy while interacting with one another), detailed measurements prove that the radiative heat is no greater than 3 degrees above absolute zero (or some −270 degrees Celsius; for comparison, water freezes at 0 degrees Celsius and boils at 100 degrees Celsius).

The cosmic background radiation is presumed to be a remnant of the fiery origins of the Universe—a Universe that has greatly cooled during the past fifteen billion years or so. In fact, this weak radiation is widely interpreted as a veritable "fossil" or "relic" of the primeval explosion that commenced the cosmic expansion long ago. Regardless of whether the initial event was a unique Big Bang causing the Universe to expand forevermore or one of several smaller bangs perhaps leading to an oscillating Universe, the primordial hot, dense matter must have emitted thermal radiation. All objects having any heat release energy; an intensely hot piece of metal (a branding iron, for instance) glows with a red- or white-hot brilliance, whereas less hot metal (such as a laundry iron) feels warm to the touch while emitting less energetic infrared and radio radiation. In its fiery beginnings the Universe almost certainly released extremely energetic gamma-ray radiation. But with time the Universe expanded, thinned, and cooled, causing the emitted radiation to shift steadily from the high-energy gamma- and X-ray varieties normally associated with superhot matter, down through the less energetic ultraviolet, visible, and infrared types, eventually to the low-energy radio variety usually released by relatively cool matter.

So, while the recession of the galaxies implies that at some past time the Universe emerged as a compact, hot, primal blob, the cosmic background radiation virtually proves it. Together these two key observations—of material objects and of radiative energy—not only strongly support an evolutionary Universe but also enable us to build meaningful models of its change throughout the course of time. These models, sometimes termed "numerical experiments," are essentially number-crunching exercises that utilize a mathematical knowledge of the laws of physics and often a super-computer.

To place the highlights of cosmic change into the grand perspective of what I like to call the "broadest view of the biggest picture," I prefer to specify the two most important thermodynamic quantities—density and temperature—at each of six major epochs in the history of the Universe. I tell my students that if physicists know the density and temperature of any object, then a great deal of information can be derived about the physical state of that object. Provided that we can both think big enough to regard the Universe as hardly different from "any object" and develop a means to derive the *average* density and *average* temperature of everything in the Universe at any moment in time, then we shall be able to appreciate the bulk temporal behavior of matter and energy throughout eternity. After all, these two properties serve to define operationally the essence of modern astrophysics: the study of the interactions of matter and energy in the cosmos.

Not surprisingly, our models of the early Universe are tentative; we say, perhaps somewhat heretically, "In the beginning there was chaos." Though we remain uncertain about creation itself (i.e., precisely the "zeroth" moment about which I speculate more in the next section), I do feel that we can effectively specify the physical conditions to well within the first second of existence—a mere moment beyond the alpha point of space and time. Admittedly, as I demonstrate in the Appendix, observations of the most distant galaxies provide direct information no closer to creation than about a billion years (i.e., about 14,000,000,000 years ago) and observations of the cosmic background radiation no closer than about a half million years (~14,999,500,000 years ago). But mathematical sketches based on a good deal of relevant supporting data and reasonable theoretical insight do yield some indirect idea of conditions almost unimaginably close to the origin of origins.

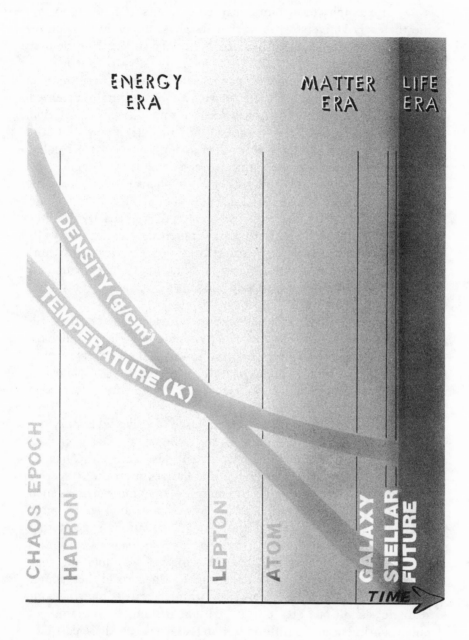

. . . then we shall be able to appreciate the bulk temporal behavior of matter and energy throughout eternity.

151

For example, most numerical experiments designed to examine the early Universe suggest that its physical conditions can be realistically specified to within a trillionth of a trillionth of the first second of existence—i.e., 10^{-24} second, or roughly the time needed for light to cross an elementary particle. (Incidentally, all the quoted time intervals in this book are referred to our culturally invented "Earth time," which is ultimately derived from our planet's rotation on its axis.) This close to creation, the currently known laws of physics specify an average density of about 10^{50} grams per cubic centimeter and an average temperature of roughly 10^{20} degrees. Of course, these huge values are hardly comprehensible to those of us living on a planet whose familiar values include, for example, one gram and eight grams per cubic centimeter for the density of water and iron, respectively, and a mere hundred degrees for the well-known temperature changes separating ice, liquid water, and steam; even the density in the Sun's core is hardly more than a hundred (10^2) grams per cubic centimeter, and its temperature "only" ten million (10^7) degrees.

Since recent research has seemingly made some startling advances in the subject of elementary particle physics, I shall have more to say about this very earliest epoch of the Universe in the next section of this chapter. For now let us continue to paint our broadest perspective and in the process encounter the first of two great changes in cosmic evolution.

The second major construction period of universal history is a bit closer to human comprehension, although it is still characterized by severely nonterrestrial conditions. Called the hadron epoch, the name of this second period derives from the collective label given the heavy, strongly interacting elementary particles such as protons, neutrons, and mesons, which were among the most abundant types of matter at the time. Calculations suggest that within a microsecond of its being, the Universe was filled with a whole mélange of such particles; whizzing this way and that, they must have haunted every available niche. Whence did they come? From energy. Under the prevailing (though still somewhat surrealistic) conditions of a million trillion trillion (10^{30}) grams per cubic centimeter and a million billion (10^{15}) degrees, these particles originated by means of a straightforward materialization—a creation—of matter from the energy of the primeval bang. No supernatural phenomena need be invoked to account for these particles, for their appearance results from the well-understood and experimentally verified physical process of "pair creation." Here elementary building blocks of matter can naturally emerge from clashes among packets of energy, in much the opposite way that a particle and its antiparticle pair are known to yield pure energy upon collision ("pair annihilation"). Hadrons doubtless interacted and collided with one another and with other types of fundamental particles, for the matter density was unimaginably large well within this first second of existence. But considering the inferno also prevalent in the Universe at this time, such particles surely remained unbound, elementary entities; the environment was just too hot for them to have assembled into anything more ordered. In such an expanding Universe the energy eventually dispersed and the creation of new hadrons from energy became progressively less likely. Accordingly, the dominant action of this epoch was the self-annihilation of hadrons into high-energy radiation, thus contributing to a brilliant fireball of mostly gamma rays, X rays, and blinding light.

As the Universe continued to expand rapidly, its contents thinned and cooled. About a millisecond after the bang, the superenergetic conditions suitable for hadron creation/annihilation had nearly subsided, allowing the lighter particles, such as electrons, neutrinos, and muons, to predominate. Thus began another process of materialization which fashioned a whole new class of elementary particles, termed leptons. By the time of this so-called lepton epoch, the average density equaled some ten billion (10^{10}) grams per cubic centimeter and the average temperature roughly ten billion degrees—physical conditions in one respect greatly moderated compared to those hugely dense and hot properties extant a fraction of a second earlier, yet in another respect still severe compared to those of our Earthly familiarity. Before the first second had elapsed, the just-created leptons were mostly self-annihilating, much like the hadrons earlier, thus fueling the radiative fireball of this cosmic bomb with still more high-energy radiation.

Of central import for us in this book, the violent conditions of the early Universe guaranteed that the energy housed in radiation greatly exceeded that effectively contained in matter. (Recall that radiation is a form of "pure" energy, whereas matter has an "equivalent" energy given by Einstein's famous formula $E = mc^2$.) Not only did the waves of radiation (or, more correctly, the radiative particles called photons) far outnumber any particles of matter, but more important, most of the energy in the early Universe took the form of radiation, not matter. As soon as the elementary particles of matter tried to assemble into something more substantive, the fierce radiation destroyed them, thus precluding the existence of even the simplest type of matter we now call "atoms," let alone any stars or galaxies. For this reason, these first three epochs are often collectively called the Radiation Era or the Energy Era. (I shall use the latter name in this book to emphasize the fact that the Universe was overwhelmingly bathed in energy during its very earliest stages, even before its newly created particles began to self-annihilate into radiation.) Whatever matter managed to exist in the Energy Era did so as a relatively thin microscopic precipitate suspended in a macroscopic sea of dense, brilliant radiation.

As time elapsed, change continued. The fourth major phase of cosmic evolution—the atom epoch—extends in time from a few minutes to about a million years after the bang. Midway through this epoch, the average density had decreased to a value of about a billionth (10^{-9}) of a gram per cubic centimeter, while the average temperature had fallen to some hundred thousand (10^5) degrees— values hardly different from those in the atmospheres of stars today. A principal feature of the atom epoch was the steady waning of the original fireball; the Universe by this time had expanded considerably, and the annihilation of hadrons and leptons had all but ceased. Even as the fireball faltered, though, a dramatic change began.

For the first few hundred centuries of the Universe, including much of the early portions of the atom epoch, energy reigned supreme over matter. All space was absolutely irradiated with light, X rays, and gamma rays, guaranteeing a structureless and highly uniform blob of radiation; we say that matter and radiation were intimately coupled to each other, in a sense equilibrated. As the universal expansion paralleled the march of time, however, the energy housed in radiation decreased faster than the energy equivalently contained in matter. (Specifically, as I show in the Appendix, the energy density of radiation decreases as the fourth power of the size of the Universe, whereas the energy density of matter decreases only as the third power.) This evolutionary imbalance ultimately caused the opaque fog of blinding radiation to lift, thus diminishing the early dominance of energy. Matter and radiation began to uncouple, their equilibrium unraveling, as a change of first magnitude set in.

Sometime between the first few millennia and a million years after the bang—an exact moment cannot be established because of the gradual nature of the change from an ionized state (known as "plasma") to a neutral state—the temperature had fallen enough to allow the charged elementary particles of matter to cluster into the more substantive atoms. (Laboratory studies on Earth have proved that at temperatures above several thousand degrees, collisions are sufficiently violent and frequent to shatter lightweight atoms; at lower temperatures, collisions are insufficient to do so.) Owing to their *charged* nature, the particles' own electromagnetic forces simply pulled them together, sporadically at first and then more frequently, since the weakened radiation could no longer split the atoms. Matter, as it neutralized, finally gained some leverage in a Universe previously ruled by pure energy. In a sense, matter had managed to overthrow the cosmic fireball while emerging as the dominant constituent of the Universe.

This phase transition from the ionized to the neutral state—an evolution from energy- to matter-dominance—signifies as fundamental a change as has ever occurred in cosmic history. I used to tell my students that this change was the greatest event of all time (save perhaps creation itself). But now having recognized another, perhaps equally significant event, to be discussed in the last section of this chapter, I judge matter's rise to dominance to be the first of two preeminent changes in all of cosmic evolution. To denote this major turn of events, most of the atom epoch and all the remaining epochs that have since transpired are collectively termed the Matter Era.

Once the Matter Era was fully established, atoms were literally everywhere. The influence of radiation had grown so weak that it could no longer prevent the attachment of the leptons and hadrons that had survived annihilation. Accordingly, hydrogen atoms were the first type of element to form, requiring only that a then slower-moving negatively charged electron be electromagnetically linked to a positively charged proton. Copious quantities of hydrogen were thus synthesized in the early Universe, and it is for this reason that we regard hydrogen as the common ancestor of all things.

By the end of the atom epoch the Universe had evolved dramat-

155

ically. The spectacularly bright fireball associated with the immediate aftermath of creation had waned; the previously opaque cosmos had become transparent. The physical conditions of temperature and density that guide all changes in the Universe had themselves undergone extraordinary change. And matter had wrested firm control from energy, thus heralding a whole new era.

Thereafter major events in the Universe were less frequent. Change continued, to be sure, but at a more relaxed pace. For once the Universe had cooled enough to allow the formation of atoms (albeit thinly dispersed ones), subsequent events would necessarily occur more slowly.

Sometime during the fifth or galaxy epoch, gravity (i.e., gravitational instabilities and statistical fluctuations) caused some of the matter to gather into vast clumps. Galaxies and galaxy clusters thereby began forming by means that we do not yet fully understand, though toward the end of the next section of this chapter I shall speculate about a currently popular idea that might explain the origin of these grandest structures in nature. One item we do know with reasonable certainty is that the quasars and remotest galaxies must have arisen in the earliest parts of this epoch; indeed, all the galaxies must have originated long ago, for observations imply that none has formed within the past ten billion years or so. Midway through the galaxy epoch, according to our computations, the average density of the Universe had decreased by another factor of more than a billion, reaching some 10^{-20} gram per cubic centimeter; the average temperature had also declined to a relatively cool few hundred degrees above absolute zero. The entire Universe was growing ever thinner, colder, and darker.

Events had slowed by this time, many billions of years after the bang. Although the early Universe was characterized by rapid change, especially in the first few minutes of the Energy Era, the later Universe changed more gradually. But it changed nonetheless.

Finally, in our sequence of six temporal intervals stretching across all time, we have the stellar epoch—the one currently engulfing us in space and time. Professional astrophysicists maintain with considerable assurance that at least ten billion years have passed since creation. In fact, the Universe could be almost twice that old; its precise age depends on the yet-to-be-determined deceleration rate of the Universe (cf. the Appendix). As the name implies, the dominant action of this epoch is the formation of stars—objects intermediate in size between atoms and galaxies. Of this we are now certain, for research during the past decade has provided direct observational evidence that stars actually are originating from the mishmash of galactic gas and dust; galaxies themselves are apparently not forming in the current epoch, but stars within them quite definitely are. At the present time the average density of the Universe is nearly a billion times thinner than in the previous galaxy epoch, approximately 10^{-30} gram per cubic centimeter. This is the critical value above which the Universe will eventually contract back to a point much like that from which it began, and below which the Universe will expand forevermore. Subtle observational tests, now under way, have not yet specified the precise density, nor therefore do we know the ultimate fate of the Universe. And as for the current temperature of everything in the Universe, galaxies, stars, and empty space alike average three degrees above absolute zero (or nearly several hundred degrees below zero Celsius). This is the cooled relic of the awesomely hot fireball prevalent eons ago, the fossilized grandeur of an ancient and glorious era. It's neither a figment of our imaginations nor an untested prediction of our Big Bang models; as noted in Chapter 1, the three-degree cosmic background radiation has been clearly detected with radio telescopes.

Mentioned here only for completeness, though I shall have more to say in the last section of this chapter, is an interesting natural by-product of the stellar epoch: the associated coagulation of matter into planets, life, and intelligence, the "mere details" of Weinberg's earlier tongue-in-cheek assertion.

> *Tiger! Tiger, burning bright*
> *In the forests of the night,*
> *What immortal hand or eye*
> *Could frame thy fearful symmetry?*
> —*William Blake*

The quest to unify all the known forces of nature has recently synthesized some aspects of the subjects of cosmology and particle physics. The electromagnetic force binding atoms and molecules and the weak nuclear force governing the decay of radioactive matter have been merged by a theory that asserts them to be different

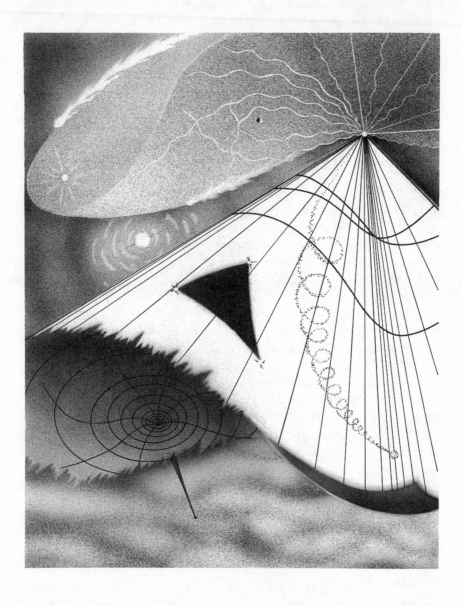

The quest to unify all the known forces of nature has recently synthesized some aspects of the subjects of cosmology and particle physics.

manifestations of one and the same force—an "electroweak" force. Crucial parts of this theory have recently been confirmed by experimenters using the world's most powerful particle accelerator in Geneva, and concerted efforts are now under way to extend this unified theory to include the strong nuclear force that binds elementary particles within nuclei. Furthermore, though we are unsure at this time how to incorporate into this comprehensive theory the fourth known force (gravity), there is good reason to suspect that we are nearing the realization of Einstein's dream: understanding all the forces of nature as different aspects of a single fundamental force.

This intellectual synthesis of the macrodomain of cosmology (for gravity is a demonstrably long-range force) and the microdomain of particle physics is but a small part of the grand scenario of cosmic evolution. Yet it is an important one, for this newly emerging interdisciplinary specialty of "particle cosmology" could provide great insight into the earliest epoch of the Universe, the time interval that I colloquially labeled "chaos" in the previous section.

Let me briefly explain the operation of the newly understood electroweak force. In microscopic (quantum) physics, forces between two elementary particles are represented by the exchange of another type of particle called a boson; in effect, the two particles can be imagined to be playing a rapid game of catch, with a boson used as a ball. In ordinary electromagnetism familiar to us in our terrestrial surroundings, the boson is called a "photon"—a bundle of electromagnetic energy that always travels at the speed of light. The new electroweak theory includes four such bosons: the usual photon, as well as three other particles having the innocuous names of W^+, W^-, and Z°. At relatively low temperatures—let us say, below about a million billion degrees, which is the range encompassing virtually everything we know about on Earth and in the stars—these bosons split into two families: the photon that expresses the usual electromagnetic force and the other three that carry the weak force. But at temperatures higher than 10^{15} degrees, these bosons work together in such a way as to make indistinguishable the weak and electromagnetic forces. Thus, by experimentally studying the behavior of this new force, we gain insight into not only the essence of nature's building blocks but also the early epochs (especially the hadron epoch) of the Universe when the temperature was indeed that high. Recalling our earlier macroscopic maxim, "Observing out into space is equivalent to probing back into time," let me now suggest a parallel phrase for the microscopic world: "Studying events at the highest temperatures (energies) is equivalent to probing the earliest moments of the Universe."

To appreciate the nature of matter at temperatures substantially higher than 10^{15} degrees and thereby to explore indirectly times even closer to creation, physicists are now researching a more general theory that incorporates the electroweak and strong nuclear forces (but not yet gravity). Several versions of this so-called grand unified theory, dubbed GUT for short, have been proposed, though experimentation capable of determining which, if any, of these theories is correct has really only begun. Like the other forces I've discussed, this grand force is mediated by a boson elementary particle, in this case called the "X boson." It is, according to these theories, the very massive (and thus energetic) X bosons that play a vital role in the first instants of time.

Imagine a time just 10^{-39} second after creation, when the temperature was some 10^{30} degrees. At that moment only one type of force other than gravitation operated: the grand unified force noted above. According to the theory of such a force, the matter of the Universe must have exerted a very high pressure that pushed outward in all directions. (Classically, pressure is the product of density and temperature, so if, in the early Universe, each of these latter quantities is large, then the pressure is guaranteed to be vast.) The Universe must have responded to this pressure by expanding in a regular way, as described in the previous section; the temperature dropped as the Universe ballooned, and in a manner inversely proportional to the size of the Universe. Thus, for example, as time advanced from 10^{-39} second to 10^{-35} second, the Universe grew another couple of orders of magnitude and the temperature fell to 10^{28} degrees.

Now, according to most grand unified theories, this temperature—10^{28} degrees—is special, for at this value a dramatic change occurs in the expansion of the Universe. In short, when matter is "cooler" than this temperature, the X bosons can no longer be produced; after 10^{-35} second, the energy needed to create such particles was too dispersed owing to the diminishing temperature. As the temperature fell below 10^{28} degrees, the disappearance of the X bosons caused a surge of energy roughly like that released as latent heat when water freezes (an event that occasionally bursts a closed container in the process). After all, energy no longer concentrated enough to yield X bosons was nonetheless available to enhance the general expansion of the Universe, in fact to cause it to expand violently or "burst" for a short duration just after the demise of the bosons. The youthful Universe, though incredibly hot, was quite definitely cooling and in this way experienced a series of such "freezings" while passing progressively toward cooler states of being. Perhaps the most impressive of all such transitions, the rather rapid decay of the X bosons caused a tremendous acceleration in the rate of expansion. This period of (actually exponential) expansion has been popularly termed "inflation"; in a mere 10^{-35} second the Universe inflated some 10^{20} times or more, smoothing out (by stretching) any irregularities existing at the outset, much as crinkles on a balloon vanish as it is inflated.

At the conclusion of the inflationary phase, some 10^{-35} second after creation, the X bosons had disappeared forever, and with them the grand unified force. In its place were the electroweak and strong nuclear forces that operate around us in our more familiar, lower temperature Universe of today. With these new forces in control (along with gravity), the Universe resumed its more leisurely expansion.

Can we test this GUT proposal, including its implied, rather spectacular inflationary change? The answer is a qualified yes, for we can do so only indirectly. After all, the world's most powerful particle accelerator has only recently created, for the briefest of instants, a temperature of 10^{15} degrees sufficient to confirm the electroweak theory. The grand unified theories become operative at temperatures only in excess of 10^{28} degrees, which physicists will likely never be able to simulate on Earth. To accelerate elementary particles to the required huge energies (that is, to effectively achieve such high temperatures in the laboratory), an accelerator machine would need to be built spanning the distance between Earth and the Alpha Centauri star system some four light-years away. (In fact, the operation of such a truly cosmic device *for merely a few seconds* would require an altogether unreasonable expenditure of power equal to the cost of several times the United States' current gross national product!) So, while physicists have successfully simulated in the laboratory the physical conditions characterizing the hadron

161

epoch, we currently have little hope of reproducing chaos. Perhaps that is for the best, as chaos is one state of nature that seems literally too hot to handle.

So how can we test the grand unified theories? One apparent success of the GUTs is that they can generally account for the observed excess of matter over antimatter. It so happens that the decay of the X bosons within the first 10^{-35} second of the Universe's being is expected to have created slightly greater numbers of protons than antiprotons. Specifically, theoretical calculations suggest that for every billion antiprotons (or 10^9 positrons), one billion and one protons (or $10^9 + 1$ electrons) were created; the billion matched pairs subsequently annihilated each other, leaving a residue of ordinary matter from which all things—including ourselves—have emerged. If this "symmetry-breaking" imbalance is true, then the matter extant today is just a tiny fraction of that formed originally.

This prediction can be tested in a straightforward way, for if protons can be created they can also be destroyed. Using the grand unified theories, we can estimate the lifetime, or average life expectancy, of the proton; it turns out to be more than a thousand billion billion times the current age of the Universe (i.e., roughly an incredible 10^{32} years)! This extremely long lifetime guarantees that although all matter is ultimately destined to disappear, the probability of decay in any given time span is exceedingly small. Nonetheless, most GUTs do suggest that protons are inherently unstable—not the immortal building blocks we once thought—and thus, any one proton is statistically in danger of decaying at any moment. In fact, since water is an abundant source of protons, theory predicts that on average one proton should decay per year in each ton of water. Alternatively expressed: A human body is expected to lose only about a single proton in an entire human lifetime. Experiments are now in progress attempting to detect such events in huge quantities of water stored in tanks in deep midwestern mines, where the water can be insulated from spurious effects caused by cosmic-ray particles reaching Earth from outer space. Furthermore, a statistical measurement of a proton's lifetime should enable us to discriminate among the various GUTs, further refining our "approximations of reality."

Another prediction of the inflationary scenario—and thus another indirect test of the many different GUTs—concerns the origin of galaxies, a subject that I claim in *Cosmic Dawn* is perhaps the greatest missing link in all cosmic evolution. Much as we discussed toward the end of the last chapter, we can reason that any, even extremely small-scale fluctuations in the matter density before inflation—and such fluctuations are an inevitable consequence of quantum physics—would be amplified by inflation to an extent now characterizing whole galaxies and clusters of galaxies. Thus the growth of gravitational instabilities, greatly aided by inflation, would have gradually led to the formation of self-gravitating collections of matter. Should this idea be correct, then, surprisingly enough, the vast agglomerations of matter we see today as galaxies, galaxy clusters, and even the truly vast galaxy superclusters are the progeny of quantum fluctuations prevalent when the Universe was a mere 10^{-35} second old. The next generation of large telescopes capable of probing the most ancient material realms of the cosmos, especially NASA's orbiting *Hubble Space Telescope*, to be launched in 1989, should soon be gathering data to elucidate the origin of galaxies and thus testing the behavior of such primeval fluctuations. Somewhat ironically, with the physicists unable to build apparatae on Earth sufficiently energetic to reproduce cosmic chaos, it is the astronomers who, by studying the macrorealm, are beginning to provide tests, albeit indirect ones, of the grand unification of the microrealm.

What about even earlier phases of the chaos epoch—earlier than 10^{-35} second? Can we probe, even theoretically, any closer to that maelstrom called creation which occurred—by definition—at time equal to zero seconds, the celebrated "t = 0" moment? Our efforts are currently hampered because doing so requires the gravitational force to be incorporated into the correct GUT. To be sure, no one has yet succeeded in developing a "supergrand" unified theory (or "super-GUT" in frontier parlance), as this is tantamount to inventing a quantum theory of gravity (or, alternatively stated, incorporating Heisenberg's Uncertainty Principle into Einstein's Relativity Theory—a prospect that was anathema to the great relativist). Even so, our current knowledge of strong gravitational forces implies that such quantum effects will likely become important whenever the Universe is even more energetic than we have yet considered. Aside from the unfathomable black holes, such huge energies could have prevailed only at times earlier than 10^{-35} second, when the Universe was even hotter and denser. Specifically, at a time of 10^{-43} second, when the average temperature was about 10^{32} degrees, the four known basic forces are thought to have been one—a truly fundamental force operating at energies characterizing the earliest part of the chaos epoch. Only at smaller energies (i.e., at times after 10^{-43} second) would the more familiar four forces begin to manifest themselves distinctly, though in reality all four are merely different aspects of the single, fundamental, supergrand force that ruled at (or near) creation.

To penetrate even closer to creation is currently hardly more than conjecture, though many researchers have a "gut feeling" that once we have in hand the proper theory of quantum gravitation, our understanding might automatically include a *natural* description of creation itself. To this end, it is not inconceivable that the primal energy emerged at zero time from quite literally nothing, uncannily in accord with the structureless singularity described by the time-honored poetic expression ". . . without form and void, with darkness upon the face of the deep. . . ." This might be true because even in a perfect vacuum—a region of space containing neither matter nor energy—particle-antiparticle pairs are constantly created and annihilated in a time span too short to observe. Though it would seem impossible that a particle could materialize out of nothing—not even from energy—it so happens that no laws of physics are violated because the particle is annihilated by its corresponding antiparticle before either one can be detected. Furthermore, for such events *not* to happen would violate the laws of quantum physics, which cite, via the Heisenberg Uncertainty Principle, the impossibility of determining *exactly* the energy content of a system at every moment in time. Hence, natural fluctuations in energy content (a little like those discussed in the previous chapter) must occur *even when the average energy present is zero.*

In this way our Universe may well have originated *ex nihilo* by means of an energy change that lasted for an unimaginably short duration—a "self-creating Universe" that erupted into existence spontaneously, much as elementary particles occasionally and suddenly originate from nowhere during certain subnuclear reactions. Such a "statistical" creation of the primal cosmic energy from absolutely nothing has been somewhat sacrilegiously dubbed the "ultimate free lunch." It may indeed be the ultimate manifestation of the long-standing quip "Nature abhors a vacuum." It might also be the solution to Leibniz's philosophical query "Why is there something instead of nothing?" The answer, ostensibly, is that the probability is greater that "something" rather than "nothing" will happen. Clearly the development of a quantum gravitational description of events at "t = 0" is the foremost challenge in the subject of physics today.

Frankly, much more so than I would have dared admit just a few years ago when I stated in *Cosmic Dawn* that the origin of the Universe is "so formidable that scientists are currently unaware even of how to go about formulating it," I now suspect that recent scientific research is destined to unsettlingly challenge philosophical and theological disciplines as well. Even so, I doubt that grand unification will grant us the means to *solve* the origin of the Universe; as numerous theologians rightly inquired of me after I gave a recent series of lectures on the subject, "How did the Universe, not yet created, know enough to obey the rules of quantum physics?" A Jesuit put it succinctly: "Whence did the laws of physics come?"

> Only spring sun, so tense, so pale
> so unestablished, could manage the matter,
> but once again as every year
> this weathered old beast
> unfolds a plan to hurl
> a filament of itself
> across the arroyo
> and build its annual bridge
> between past and future.
> If I stand back, keep
> an eye on things, I may
> witness my own delivery
> from high above the female proscenium,
> cave of fertility and its enactment.
> This event is me. Do I know it?
> Am I this very thing?
> I answer my own questions.
> An intimate galaxy of genes
> reiterates the permanence of change,
> and this great attic of language,
> lovingly rummaged,
> will welcome a fresh brood of tenants
> into the most tumbledown of houses.
>
> Let us live here.
>   —Peter Davison

Of all the known clumps of matter in the Universe, life forms, especially those enjoying membership in advanced technological civilizations, are arguably the most fascinating. What's more, I suggest that technologically competent life differs fundamentally from lower forms of life and from other types of matter scattered throughout the Universe. This is hardly an anthropocentric statement; I say this because after more than ten billion years of cosmic evolution, the dominant species on planet Earth—namely, we human beings—has learned to tinker not only with matter but also with evolution. Whereas previously the gene (i.e., DNA) and the environment (whether stellar, planetary, geological, or cultural) governed evolution, twentieth-century Earthlings are rather suddenly gaining control of aspects of both these agents of change. We are now tampering with matter, diminishing the resources of our planet while constructing the trappings of utility and comfort. And we now stand at the verge of manipulating life itself, potentially altering the genetic makeup of human beings. The physicist unleashes the forces of nature; the biologist experiments with the structure of genes; the psychologist influences behavior with drugs. We are, in fact, forcing a change in the way things change.

The emergence of technologically intelligent life, on Earth and perhaps elsewhere, heralds a whole new era: a *Life Era*. Why? Because technology, for all its pitfalls, enables life to begin to control matter, much as matter evolved to control radiative energy more than ten billion years ago. As such, matter is now losing its total dominance, at least at those isolated residences of technological competence. To use a cliché that is true as never before, life is now taking matter into its own hands.

A central question before us is this: How did the neural network within human beings grow to the complexity needed to fashion societies, weapons, cathedrals, philosophies, and the like? To appreciate the essence of life's development, especially of life's evolving dominance, we return to some of the thermodynamic issues raised in the previous chapter.

Recall the hot and dense conditions prevalent in the early Universe. Until such time that neutral atoms began forming—the recombination phase change some half million years after creation—radiation and matter were intimately coupled. Equilibrium obtained during the Energy Era as a single temperature was enough to describe both radiation and matter. In the previous chapter I characterized such a state by a lack of order or, as it is more commonly phrased, by maximum entropy. (Note that the entropy referred to here is the thermal or chemical entropy, which is indeed maximized, not the gravitational entropy, which is not at all high in these earliest epochs.)

According to our earlier arguments regarding equilibrium thermodynamics, a lack of order or structure implies a lack of information. The absence of a temperature gradient between radiation and matter in fact guaranteed zero information (or zero macroscopic order) in the early Universe. This is the basis for the chaotic state at the start of the Energy Era that I discussed earlier in this chapter. Once radiation and matter decoupled, however, equilibrium was destroyed, a thermal gradient developed, and the Matter Era became established. Two temperatures were thereafter needed to describe the evolution of radiation and matter. Whereas the "radiation temperature" continued to decline at the same rate characterizing change during the early Universe, the "matter temperature" decreased more rapidly. (As derived in the Appendix, the latter varies inversely as the time, whereas the former varies inversely as the square root of the time.)

Thus, the very expansion of the Universe drives order from chaos; the process of cosmic evolution itself generates information. How that order became manifest in the guise of galaxies, stars, planets, and life forms has not yet been deciphered in detail. But we can now identify the essence of the development of natural macroscopic systems—ordered physical structures able to assimilate and maintain information by means of local reductions in entropy—in a Universe that was previously unstructured in the extreme.

Furthermore, because the two temperatures characterizing the Matter Era diverge—that is, their difference becomes larger with time—the growing departure from thermodynamic equilibrium allows the cosmos to produce *increasing* amounts of negentropy or macroscopic information. We thereby seemingly have a means to appreciate, at least in gross fashion, the observed rise in complexity throughout the eons of cosmic evolution. Indeed, we shall need to identify a way of generating substantial amounts of order, negentropy, and information if we are to justify the emergence of structures as complex and intricate as a single cell, let alone the neural architecture of the human brain.

The growth of information throughout cosmic history has not

been steady; rising slowly in the early parts of the Matter Era, it has climbed more rapidly in recent times. This temporal increase in macroscopic order runs completely opposite to the decline of the density of energy contained in both radiation and matter. Whereas the previously discussed eclipse of radiation energy density by matter energy density signified the onset of the Matter Era, the ascent of information density over both these energy densities heralds the Life Era. In particular, the controlled use of radiation by animate structures (i.e., when the information density exceeds the radiation energy density, beginning with the simplest type of autotrophic cell) may be identified as one of the first great inventions of biological evolution: photosynthesis. This is a grand event wherein life dominates radiation, utilizing starlight in a survival-related fashion. An even grander event occurs when life forms dominate matter (i.e., when the information density exceeds the matter energy density, beginning with technological intelligence on Earth and possibly elsewhere). Only the latter of these two changes heralds a genuinely new era because only with the origin of technologically manipulative life, not just life itself, does life exert leverage over *both* radiation and matter.

Cosmic expansion is not the only source of structural order in the Universe (though it does seem to be a necessary, if not sufficient, condition). On local scales the evolution of gravitationally bound systems, for example, can also generate information. An ordinary star is a good case in point. Such astronomical objects are now well known to originate from dense pockets of gas and dust within otherwise chemically and thermally homogeneous galactic clouds. Initially the young star has only a relatively small temperature gradient from core to surface, and is normally composed of an approximately uniform mixture of ninety percent hydrogen and ten percent helium nuclei, often peppered with trace amounts of heavier elemental nuclei. As the star evolves, its core progressively increases in temperature while adjusting its size like a cosmic thermostat; all the while nuclear fusion reactions change its lightweight elements into heavier types. With time, then, such an object grows thermally and chemically inhomogeneous, gradually becoming more ordered and less equilibrated. While enhancing the structure

. . . such an object grows thermally and chemically inhomogeneous, gradually becoming more ordered and less equilibrated.

of a star, such stellar evolutionary processes inevitably generate and store information, for a complete description of a thermally and chemically differentiated system requires more data than an equally complete description of its initially homogeneous state.

Thus, even though stars are energy (and entropy) converters, they represent relatively localized sites of growing information, first, because stars also *radiate* entropy (as well as energy) into their surrounding environment and, second, because the gravitational agents tending to enhance stellar gradients usually overwhelm the opposing (nongravitational) agents tending to diminish them. In essence, stars are islands of increasing negentropy.

Consider a final point regarding open, self-gravitating systems—which, by the way, stars are. This kind of a "gas without walls" is quite unlike an isolated system in a laboratory. Stars, among other members of the cosmic hierarchy of material clusters, do not relax toward a state of thermodynamic equilibrium with their surroundings. Even the terminal phase of many stars is not at all an equilibrium state. Instead, their ultimate fate is one of free gravitational infall. When the star loses even its (previously maintained) hydrostatic equilibrium between inward-pulling gravity and outward-pressing gas, it collapses catastrophically toward a bizarre configuration known as a black hole.

But enough about inanimate objects. What about living systems, especially their attendant growth of biological structure? Surely, negentropy must increase during life's origin and evolution, for living systems are demonstrable storehouses of concentrated energy, low entropy, and considerable order. As with other objects in the Universe, we can use information theory to describe both the structural and the functional aspects of biological organization.

All things considered, biological systems are best characterized by their coherent behavior, for their maintenance of order requires a great number of metabolizing and synthesizing chemical reactions as well as a host of complex mechanisms controlling the rate and timing of many varied processes. But this doesn't mean that life violates the second law of thermodynamics, a popular misconception. Although living organisms manage to decrease entropy locally, they do so at the expense of their environment—in short, by increasing the overall entropy of the remaining Universe.

Living things are able to "circumvent" temporarily the normal entropy process by absorbing available (free) energy from their surrounding environment. They do so—during both their origin and their evolution—because of temperature gradients in Earth's atmosphere. What is the origin of these thermal differences and ultimately of the energy utilized in the process of living? On Earth, it's our Sun. Energy flows from the hot (six-thousand-degree-Celsius) surface of the Sun to our relatively cool (twenty-five-degree-Celsius) planet. Of crucial importance, useful work can be done with this available energy.

All plants and animals depend on the Sun for survival. Green plants photosynthesize by using direct sunlight to convert water and carbon dioxide into nourishing carbohydrates; animals obtain the Sun's free energy more indirectly by eating plants and other animals. In his seminal compilation of lectures *What Is Life?*, the Austrian quantum physicist Erwin Schrödinger expressed it succinctly: An organism "keeps aloof from the dangerous state of maximum entropy by continually drawing from its environment negative entropy."

Of course, if left alone, all living things, much like everything else in nature, tend toward equilibrium. While just twitching a finger (or merely thinking while reading this book without twitching a finger), we expend some energy. Any action taken indefinitely, without further energizing, would drive us toward an equilibrium state of total chaos or orderlessness. By contrast, living systems stay alive by steadily maintaining themselves far from equilibrium. They do so by means of a flow of energy through their bodies. In point of fact, unachieved equilibrium can be taken as an essential premise, even an operational definition, of all life.

As humans, for example, we maintain a reasonably comfortable steady-state by feeding off our surrounding energy sources, principally plants and other animals. I say "steady-state" since, as noted earlier for any open system, by regulating the rate of incoming energy and outgoing wastes, we can achieve a kind of stability—at least in the sense that while alive, we remain out of equilibrium by a roughly constant amount. In a paradoxical juxtaposition of terms, we might therefore describe ourselves as "dynamic steady-states." Of course, we waste much of the incoming energy while radiating heat into the environment; warm-blooded life forms are generally warmer than the surrounding air. But some of the absorbed energy can power useful work. Once this available energy flow ceases, the dynamic steady-state is abandoned, and we drift toward the more common, "static" steady-state known as death, where, following complete decay, our bodies reach a true equilibrium.

Here's what happens in the food chain consisting of grass, grasshoppers, frogs, trout, and humans. According to the second law, some available energy is turned into unavailable energy at each stage of the food chain, thus causing greater disorder in the environment. At each step of the process, when the grasshopper eats the grass, the frog eats the grasshopper, the trout eats the frog, and so on, useful energy is lost. The numbers of each species required for the next higher species to continue decreasing entropy are staggering. According to the chemist G. Tyler Miller, the support of one human for a year requires some three hundred trout. These trout, in turn, must consume ninety thousand frogs, which yet in turn devour twenty-seven million grasshoppers, which live off some thousand tons of grass. Thus, for a single human being to remain "ordered" (i.e., to live) over the course of a single year, we need the energy equivalent of tens of millions of grasshoppers or a thousand tons of grass. Clearly, then, we maintain order at the expense of an increasingly disordered environment. Every living thing, in fact, takes a toll on the environment. The only reason that the environment doesn't decay to an equilibrium state is that the Sun continues to shine daily. Indeed, the whole biosphere comprises a nonequilibrium state that is subject to solar heating, thus rendering much environmental energy. Earth's thin outer skin is thereby enriched, permitting us and other organisms to go about our business of living.

In a penetrating essay entitled "The Music of *This* Sphere," the medical doctor Lewis Thomas summed it up nicely with his characteristic felicity of style:

> [The biophysicist H. J.] Morowitz has presented the case, in thermodynamic terms, for the hypothesis that a steady flow of energy from the inexhaustible source of the sun to the unfillable sink of outer space, by way of the earth, is mathematically destined to cause the organization of matter into an increasingly ordered state. The resulting balancing act involves a ceaseless clustering of bonded atoms into molecules of higher and higher complexity, and the emergence of cycles for the storage and release of energy. In a nonequilibrium steady state, which is

postulated, the solar energy would not just flow to the earth and radiate away; it is thermodynamically inevitable that it must rearrange matter into symmetry, away from probability, against entropy, lifting it, so to speak, into a constantly changing condition of rearrangement and molecular ornamentation. In such a system, the outcome is a chancy kind of order, always on the verge of descending into chaos, held taut against probability by the unremitting, constant surge of energy from the sun.

It is instructive to pursue this point a bit further. Suppose Earth's atmosphere and outer space were to achieve thermal equilibrium. All energy flow into and out of Earth would cease, causing all thermodynamic processes on our planet to decay within surprisingly short periods of time. For example, a rough estimate shows that the reservoir of Earth's atmospheric thermal energy would become depleted within a few months, the latent heat bound in our planet's oceans would dissipate in a couple of weeks, and any mechanical energy (such as atmospheric circulation that contributes to weather as we know it) would be damped in a few days. So be sure to place Earth's energy budget into perspective; neither our planet's primary source of energy nor its ultimate sink is located on Earth.

Not only is life, at any given moment, a reservoir of order, but evolution itself also seems to foster the emergence of greater amounts of order from disorder. During the course of biological evolution, each succeeding (and successful) species becomes more complex and thus better equipped to capture and utilize available energy. The higher the species in the chain, the greater the energy density fluxing through (or the information density of) that species, and the greater the disorder created in the Earth-Sun environment. Alas, the principal source of our available energy, the Sun, is itself running down as it "pollutes" interplanetary space with increasing entropy. So use caution while regarding evolution as progress. Evolution means creation of ever more complex islands of order at the expense of even greater seas of disorder elsewhere in the Solar System as well as in the Universe beyond.

Let us be clear that the thermal gradients needed for an energy flow in Earth's biosphere could not be maintained without the Sun's converting gravitational and nuclear energies into radiation that emanates outward into unsaturable space. Were outer space ever to become saturated with radiation, all temperature gradients would necessarily vanish, and life among many other ordered structures

. . . *neither our planet's primary source of energy nor its ultimate sink is located on Earth.*

would cease to exist; this is essentially a variation of Olbers' paradox, a statement inquiring, in view of the myriad stars in the heavens, why the nighttime sky is not brightly aglow. That space will never become saturated (or the nighttime sky bright) can be attributed to the expansion of the Universe, thus bolstering the suggestion that the dynamic evolution of the cosmos is an essential condition for the order and maintenance of all things—including, at least on Earth, the origin and evolution of life itself. This gives us all the more reason to include life within our cosmic evolutionary cosmology, for the observer in the small and the Universe in the large are not disconnected.

The transition from the Matter Era to the Life Era will not be instantaneous; it is a demonstrably evolutionary, not revolutionary, process. Just as a great deal of time was needed for matter to conquer radiation in the early Universe, long durations will likely be required for life to best matter. Life may not, in fact, ever fully overwhelm matter, either because civilizations might never gain control of material resources on a truly galactic scale or because the longevity of technological civilizations everywhere might be inherently small.

Though a mature Life Era may never come to pass, one thing seems certain: Our generation on planet Earth, as well as any other neophyte technological life forms populating the Universe, is now participating in an astronomically significant transformation, a second great change in the history of the Universe. We now perceive the dawn of a whole new reign of cosmic development—an era of opportunity for life forms to begin truly to fathom their role in the cosmos, to unlock secrets of the Universe, indeed to decipher who we really are and whence we came. We have become smart enough to reflect back upon the material contents that gave us life. The implications of our newly gained power over matter are nothing short of cosmic.

CHAPTER 5

Implications for the Life Era

CONTINUED GROWTH VIA COSMIC EVOLUTION ASSERTS THE NEED FOR A GLOBAL, EVEN COSMIC SET OF ETHICS IF TECHNOLOGICALLY ADVANCED BEINGS ARE TO ENTER THE LIFE ERA

EVOLUTION, ENERGY, AND ENTROPY constantly interplay throughout my current research. As time irreversibly moves forward, each of these quantities changes, though often in peculiar ways. Particularly intriguing is the *rate* at which various things change. Accordingly, we are now using the tools of technology to explore various rates of change within and among material substances. For example, in recent years I have used telescopic equipment to probe the extraordinarily slow rates at which stars form, as well as their rates of evolution. Likewise, my bioscience colleagues employ microscopic gear to study the comparatively faster rates at which genetic mutations force changes within life forms, including the speciation of whole new organisms. To be sure, much of today's pioneering scientific research focuses not merely on what things change but also on how and when they change.

Similarly, the rapid rate at which human civilization now evolves fascinates and concerns us. I don't mean anatomical change, but technological change. Significantly, much of this rapid change is of our own making, for technology is a natural (and possibly inevitable) component of cultural evolution. As time unfolds, intelligent beings on any planet are destined to gain the ability to prescribe, inter alia, environmental, political, social, and economic change. Accordingly, the exercise of free will now determines (at least partly) the ways and means that our society utilizes energy and thus the speed with which our civilization changes. We have become technologically smart (though not necessarily wise) enough to be able to regulate change: to speed it up, to slow it down, or perhaps to stop it altogether. In short, the most advanced terrestrial life forms are now partly usurping the role previously played by inanimate nature, for we humans have become the agents of change—at least on planet Earth.

Whether we increase or decrease change or try to halt it completely, we must carefully examine the worldly consequences of change in light of our mushrooming predicament on our rather fragile planet. Only by thinking more broadly than in the past can we act wisely in the future—and by "wisely" I mean simply and clearly in the best interests of *all* humankind. Thus, I contend that our newly emerging scientific philosophy must include the study of ethics, and by implication the humanistic subjects of religion and classical philosophy as well as the "soft" sciences such as economics, law, politics, and perhaps even sociology. All this is part of cosmic evolution; such a transdisciplinary, holistic vision is truly our new philosophy's forte.

> *Praise to You, harsh Matter . . . dangerous*
> *Matter . . . which one day will be dissolved with us*
> *and carry us into the Heart of Reality. . . .*
> *Battering us and dressing our wounds . . .*
> *struggling before You yield to us . . . destroying and*
> *building, smashing and liberating. . . .*
> *Seed of our souls, Hand of God . . .*
> *I bless You.*   —*Pierre Teilhard de Chardin*

177

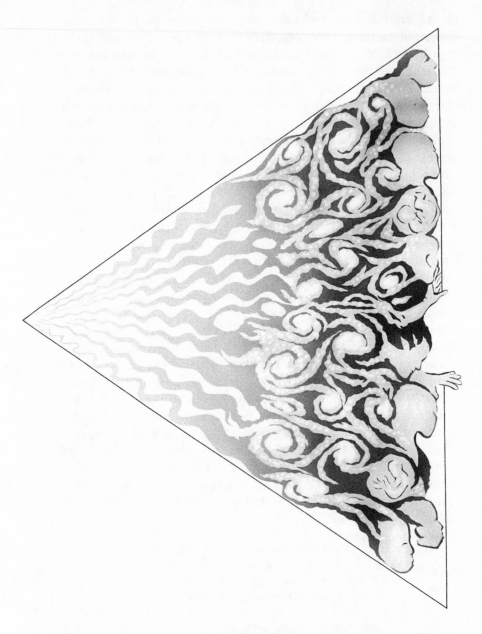

. . . the most advanced terrestrial life forms are now partly usurping the role previously played by inanimate nature, for we humans have become the agents of change . . .

While lecturing around the country on the subject of cosmic evolution during the past decade or so, I have been repeatedly asked how this evolutionary synthesis, and especially my view of the Life Era, compare with the idea advocated a half century ago by the French scientist-theologian Pierre Teilhard de Chardin. Admittedly not an aficionado of the history of science or of philosophy, I was quite unaware of Teilhard's ideology and at first begged ignorance. Yet, after one particularly insistent challenge from a New York City clergyman who implied that I had merely scientifically substantiated the Teilhardian equivalent of the Life Era, I returned to Harvard Square to do some extracurricular research. Having found the central university libraries too vast for easy reference during my student and faculty days at Harvard, I undertook much of my nonscientific studying in the Square bookstores, and my intention that day was to spend a few moments determining just who this Teilhard fellow was. However, after spending several fascinating hours reading his major work, *The Phenomenon of Man* (1955), I returned to the Observatory having purchased virtually all his collected writings.

Pierre Teilhard de Chardin (1881–1955), at the same time a Jesuit priest and a distinguished geopaleontologist, was one of the most synthesizing thinkers of this century. Forsaking the Thomist philosophy of his church and rejecting the Aristotelian dichotomy between matter and mind (or body and soul), he envisaged the whole of knowable reality as a process—a dynamic, evolutionary process encompassing not only energy, matter, and life but "spirit" as well. Doubtless he sought to include the latter in deference to his Christian calling, thereby achieving a personal union of science and religion, for he held that "Like the meridians as they approach the poles, science, philosophy and religion are bound to converge as they draw nearer to the whole." (He stressed that he did not use the verb "to merge.") Not surprisingly, however, given today's emphasis on excessive specialization in most sciences and unyielding dogmatism in most religions, Teilhard has been largely dismissed as a veritable mystic by scientists and a virtual heretic by theologians. During his own lifetime, in fact, he was censored by the Roman Curia, prohibited from teaching in France, and effectively exiled to China, where he spent nearly a quarter century, while his major works were barred from publication until after his death. Sir Julian Huxley summed up smartly the ensuing controversy in his introduction to the English edition of Teilhard's magnum opus:

> The biologist may perhaps consider that in *The Phenomenon of Man* he paid insufficient attention to genetics and the possibilities and limitations of natural selection, the theologian that his treatment of the problems of sin and suffering was inadequate or at least unorthodox, the social scientist that he failed to take sufficient account of the facts of political and social history. But he saw that what was needed at the moment was a broad sweep and a comprehensive treatment.

Unquestionably, grand synthesizers and big thinkers are even more eagerly needed (though not necessarily more welcome) today.

Much of Teilhard's thesis accords generally with our current scenario of cosmic evolution. He strives to demonstrate that any individual is related to humankind as a whole, that humankind in turn relates to all life, and, further, that life shares common ground with the Universe at large. In addition, he adopts an evolutionary viewpoint while attempting to establish a coherent order of antecedents and consequents surrounding centrally positioned humans in his grand scheme of nature. In Teilhard's words, "Thence stems the basic plan of this work: *Pre-Life: Life: Thought*—three events sketching in the past and determining for the future *(Survival)* a single and continuing trajectory, the curve of the phenomenon of man." His epistemology is absolutely prophetic of what I seek in today's cosmic evolutionary synthesis—to wit, "The true physics is that which will, one day, achieve the inclusion of man in his wholeness in a coherent picture of the world."

But differences—major and telling differences—abound between Teilhard's grand vision and our current attempts to generate the scientific philosophy of cosmic evolution. In the first sentence of his preface he claims that *The Phenomenon of Man* should be read as a work neither of metaphysics nor of theology but as "purely and simply a scientific treatise" (though I suspect he may have stated such in an unsuccessful bid to placate papal authorities). That this is clearly not a work of strictly natural science can be gleaned from his overabundance of anthropocentric statements ("Man is the *center of construction* of the Universe" [his italics]) as well as his seemingly teleological insight (". . . how can one fail to recognize

180

this revealing association of technical organization and inward spiritual concentration as the work of the same great force . . . the very same force which brought us into being?").

While I can neither disprove the latter nor offer objective refutation of the former, Teilhard's embrace of clearly spiritual elements under the guise of science calls to mind current, wholly unwarranted attempts by religious fundamentalists (almost exclusively in the United States) to gain legislative footholds in the scientific world (though in no way was Teilhard a fundamentalist; indeed, had he been one, an inquisitive Curia might not have caused him so much grief). Incidentally, this is not a criticism of fundamentalism per se as much as a statement that such should be clearly labeled religion. To confuse science and religion—in particular, to demand dogmatically a literal discussion of the Bible in science classrooms—is a serious mistake. It's even dangerous, for we would soon witness a technological retrenchment of America in the community of nations.

The most striking parallel between current expositions of cosmic evolution and that of Teilhard's religioscientific synthesis—indeed the one that almost invariably emanates from even the most passive audiences I've addressed—concerns the future orientation of humanity. Whereas I have described in this present work my concept of the Life Era, Teilhard has also written of the destiny of human evolution. He calls his newly emerging era the "noosphere," a rather bizarre term he coined (but failed to define apart from calling it a "thinking skin") to denote the domain of mind that is superposed on the sphere of life or the biosphere of Earth. Apparently, as Huxley inferred, the noosphere is "the union of the whole human species into a single inter-thinking group"—a sort of planetary organism having a personal goal which Teilhard described thus: "Everything that is hard, crusty, or rebellious . . . all that is false and reprehensible . . . all that is physically or morally evil will disappear. . . ." All of which to me sounds reasonable in principle, until he feels obliged to add that consequently "Matter will be absorbed into Spirit," a rather flat-footed claim that leaves me wanting.

Aside from issues of spiritualism, on which natural science is frankly (and appropriately) mute, the concept of "convergence" is where I most strongly differ with Teilhard. He maintains that history's evolutionary events have converged on man, who is the final product of such events—the "summit of an anthropogenesis which is itself the crown of a cosmogenesis"—a vision that I find, quite simply, self-centered in the extreme. Even so, he does approve of Nietzsche's metaphysical view that man is unfinished and will yet transcend to some higher plane of future organization—a process in classic Teilhard-speak of "ultrahominization" or "Christogenesis" that seemingly molds human life in the image of the Jesuit's God. This ultimate state of human convergence, which to some resembles an emerging divinity, he calls the Omega Point—a final condition that apparently comprises both a purpose and an end to the human adventure. By contrast, I have in no way whatever asserted or implied that humans are the pinnacle of cosmic evolution.

Be assured, I am not criticizing attempts to yoke science and religion. As a member of the Editorial Board of *Zygon: The Journal of Religion and Science*, I welcome such endeavors. Each of these ancient institutions should have meaningful roles to play in society, one to unlock secrets of the Universe and the other to provide a sense of ethics to help determine *how* we can best decipher those secrets for the betterment of all humankind (though frankly I am worried about our ability to achieve the latter). Accordingly, I often offer my students the following analogy: At the summit of a mountain lies the truth (though I am quite unprepared to define that elusive term here). Up one side of the mountain crawl the cosmologists; up the other, the theologians. As each of us gathers new information and insight through our respective methods of scholarship, we progress toward the summit. And although I doubt either culture—science or religion—will ever reach the top and thus secure the "truth," the important point is that the distance between us grows smaller all the while we ascend. In very general terms, these two great cultural edifices do seem to be converging without necessarily ever merging per se.

In any case, I regard it important to clarify that despite Teilhard's disavowal, his ideology is substantially weighted with decidedly theological overtones. "Psychic energy," "Christogenesis," and "single soul," among a host of other newly stylized Theilhardian terms, are simply not to be found in the current lexicon of natural science; for them to be acceptable to science probably *would* require a merger of science and religion. This most enlightened Jesuit, whom I personally admire, undeniably attempted to reconcile the supernatural elements of Christianity with the facts and implications of modern evolutionary theory. I, on the other hand, have endeavored to remain squarely within the confines of accepted natural science, and as such I conclude, without modesty or apology, that the Life Era I write of here is more natural, the modern scenario of cosmic evolution more objective.

Teilhard de Chardin's cosmic mysticism resembles that of the first transalpine Renaissance philosopher, Cardinal Nicholas of Cusa (1401–1464). In offering a rather early departure from the Aristotelian static worldview, Nicholas boldly hypothesized a moral cosmology by blending his interests in mathematics and mysticism. The result claimed that humans are an existential midpoint between God as the universal macrocosm and Christ as the universal microcosm.

Though Teilhard never acknowledged it, his poetic and inspired writings were also doubtlessly influenced by those of perhaps the greatest modern French philosopher. Henri Bergson (1859–1941) in 1907 had authored a seminal work, *Creative Evolution*, which postulated the organization of ordered structures by means of an esoteric vitalism, or the élan vital. Resembling Teilhard's "psychic energy" and even the "free energy" of our modern scenario of cosmic evolution, the élan vital played a pivotal role in Bergson's dualistic philosophy, wherein the world divides into two separate parts: matter and life. The whole Universe, according to Bergson, amounts to a clash and conflict between these two properties, which he viewed as opposite motions or changes: life, which is upwardly mobile, and matter, which is at best static and usually downward falling. Over the course of time, he argued, life struggles against the resistance of matter, gradually learning to use matter to further organize itself. Throughout, though subdued by the adaptations that matter forces on it, life retains its capacity for free activity, forever seeking greater liberty amidst the opposing constraints of matter. The resulting theory of creative evolution, contrary to almost all other vitalisms, predicates no ultimate goal; Bergson held that evolution diverges, in contrast with Teilhard's convergence toward a final unity. Here, then, is the modern philosophical (though hardly scientific) root of the Life Era—a period in cosmic history, following a titanic battle

"Form is only a snapshot view of a transition."

between matter and life, wherein life gains an ever greater, even controlling influence in nature.

The concept of change is uppermost in Bergson's philosophy. Mimicking early thinkers of antiquity, he reasoned that reality is not made of separate things, rather is just an endless stream of becoming. As with Heraclitus and others before him, Bergson denied the existence of things: "Form is only a snapshot view of a transition." Much to my liking, he described evolution as truly creative, akin to the work of an artist—hence the title of his Nobel Prize–winning work of literature, *Creative Evolution*. As an agnostic, though, I find that Bergson disappoints me by asserting, like Teilhard, that the evolutionary process is necessarily guided by divine purpose.

In the last analysis, Bergsonianism was a bold attempt to reconcile subjectivism and mysticism with rational thought and objective science. He gave priority, however, to the former (especially to the concept of intuition), much again as Teilhard, in his later alleged reconciliation of religion and science, often opted for religion. Bergson's philosophy is grounded in religious beliefs and metaphysical assumptions not subject to empirical verification or rational inquiry. Even so, the major contribution of both these grand synthesizers is their recognition that the fact of evolution must be seriously incorporated henceforth into all viable world systems.

The bare essence of the Life Era concept may well extend back to antiquity, Aristotle having stated, "If human beings could be shaped by their environment, they could change themselves in equal measure by their own efforts." Recall that nature in Aristotle's metaphysics is twofold, consisting of matter and form. Derived from his scheme of biology, living systems are theorized to display some "form" that emerges from matter or raw material. Formless matter is "no-thing," whereas form, more than just shape, embodies an inner need or impulse—Aristotle's internal perfecting principle noted in Chapter 2—that molds matter into specific configurations. I suspect that the Life Era's most elementary roots reside here since according to Aristotle, nature is the constant conquest of matter by form, as the Universe and all its components develop toward something "better" than what existed before, ultimately culminating in the intelligent life that we humans now share. (Not that we are perfect form; only God enjoys such, said Aristotle.) Of course, all such Aristotelian arguments, even ignoring charges of egocentrism, are not only metaphysical but distinctly teleological.

In closing this section, let me emphasize once more my claim of intellectual advantage in that the Life Era concept originates from a modern study of cosmic evolution; as essayed in this work, it is firmly grounded in natural science and makes no appeal to untested psychic intuition, untestable spirits, or imaginative teleology. Despite its movement away from strict determinism, the world of science today is decidedly materialistic, based on facts and a series of logical reasonings that follow from those facts. Admittedly my notion of this third great phase of cosmic history right now has philosophical and even religious implications, but I cannot overly stress that the *origin* of my concept of the Life Era derives solely from a naturalistic analysis of universal change as sketched in Chapters 3 and 4.

> *I sit at my desk now*
> *like a tiny proprietor,*
> *a cottage industry in every cell.*
> *Diversity is my middle name.*
> *My blood runs laps;*
> *I doubt yours does,*
> *but we share an abstract fever*
> *called thought,*
> *a common swelter of a sun.*
> *So, Beast, pause a moment,*
> *you are welcome here.*
> *I am life, and life loves life.*
> *—Diane Ackerman*

Except for a few qualifying phrases earlier in this book, I have thus far restricted my discussion of the Life Era to the role of technologically competent life forms on Earth; not that humanity has yet entered such an era, merely that we are now perched on its threshold. Since I have scorned anthropocentric arguments elsewhere, I wish to do so now concerning the Life Era. My point is

186

that should advanced aliens exist on other worlds, then each of them, by their presence alone, will contribute to the development of a cosmic Life Era, and they may well do so at different times in the Universe as each evolves the technical ability to manipulate matter on planetary and larger scales.

Recognize that I am not concerned in this section with pre-technological, creepy-crawly life forms. With respect to the Life Era, any society incapable of, let us say for the sake of argument, designing and operating a radio telescope is currently inconsequential. Neither facetious nor arrogant (nor does that statement have much to do with the fact that my training is largely in the field of radio-frequency physics), I am simply confining my analysis to those organisms possessing a certain minimum high-tech capability. Nor am I arguing that extraterrestrial intelligent life must in fact exist anywhere in the cosmos, only that it might, and I want to explore here its implications for the Life Era. Frankly I confess ignorance regarding the plurality of extraterrestrials; indeed, it is difficult to take a stand on any subject having virtually zero data. I do maintain, as I made clear more fully in *Cosmic Dawn*, that we as an advanced civilization should mount a moderate search for extraterrestrials, if only as a means of enhancing our curiosity, a curiosity which I take to be, above all else, the hallmark of a humane, vibrant, survivable society. To do otherwise, to fail to search, is tantamount to committing the cardinal sin of pre-Renaissance workers: philosophizing and pontificating without experimentally testing.

If we are now alone as an advanced life form in the Universe, then we humans are contributing uniquely to the development of a distinctly new realm in the history of the cosmos—a sobering notion to be sure, as that makes us currently the sole conscious trustee of further evolutionary progress. Should we, furthermore, be the *first* such intelligence to have emerged in the cosmos, then we are the vanguard of the Life Era—an even weightier notion. I draw this distinction because while we might well currently be the only intelligence in the Universe, it does not necessarily follow that we are the first. To see why, consider some pertinent time scales. The Universe is roughly fifteen billion years old; Earth, but one-third that age. Accordingly, some ten billion years transpired before Earth even formed as a planet. Now, if in keeping with the best scientific estimates we take the galaxies to be of the order of twelve billion

187

years old, and if the oldest stars therein require a couple of billion years to synthesize and expel a significant amount of heavy elements from which solid planets and hence life are capable of forming, then the Universe could conceivably be populated with habitable planets as old as ten billion years or double the age of Earth. Furthermore, if evolutionary progress even remotely resembles the history of life's development on Earth—roughly a billion years needed to originate life and another several billion years to develop technological intelligence—then this sort of loose order-of-magnitude reasoning implies that intelligence could have conceivably emerged on one or more planets several billion years ago. Thus, the transition to the Life Era may well have been entered billions of years ago, though admittedly I regard this as unlikely on both observational and theoretical grounds, as I shall note in a moment.

As regards the distinction I made above between being "alone" and being "first" among intelligences, we on Earth could indeed be alone if each and every one of any earlier hypothetical alien intelligences failed to survive—a proposition not entirely unreasonable, for no civilization may be able to check its tendency to self-destruct over sufficiently long time scales. At virtually any given moment in cosmic history the Universe might well be populated by a single cosmic intelligence and yet not have it be the first such intelligence. A useful analogy is in order: Imagine an ornate chandelier having a huge number of light bulbs. The chandelier could represent a finite cosmos, or more reasonably a galaxy, while the bulbs denote planets ecologically suited for the emergence of technological intelligence. Each bulb illuminates only when technology on a given planet surpasses some crucial threshold—such as, once again, radio communicative ability. Two factors generally determine whether the chandelier is blazing or dim: One concerns the vitality of the evolutionary process leading to technological life—the extent to which evolution's hand twists far enough to screw in and light a bulb; while the other concerns the length of time each bulb stays lit—the longevity of a technical civilization. At the two extremes the chandelier could be brightly lit with many glowing bulbs, indicating much intelligent activity on many planets, or it could be completely unlit, designating a technocultural void. Interestingly enough, the peculiarly popular intermediate case noted above ("alone yet not first") implies that at any given time only a single bulb is lit. Considering the many varied and speculative time scales affecting an advanced civilization—especially those for technological emergence and longevity—all the bulbs might eventually glow without any two bulbs' ever being lit simultaneously. This latter case would be especially true if the chandelier were outfitted with flashbulbs, implying that every galactic civilization is relegated to cosmic loneliness—a scenario, incidentally, prohibiting much progress toward a genuine Life Era since no one civilization endures long enough to control matter over truly galactic dimensions.

Special cases aside, I regard more likely—if only guided once again by my antianthropocentrism—the possibility that we are neither alone nor the vanguard of universal intelligence. This "assessment of mediocrity" derives largely from the assumed universality of the principles of cosmic evolution—a monolithic assumption which, if incorrect, means not only that 2 + 2 does not equal 4 everywhere but also that much of what I say in this section is untenable. Nonetheless, this assumption holds that the laws of physics pertain equally to every nook and cranny of the cosmos and is so strongly embraced by the scientific community as to be considered virtual dogma, for no natural scientist of my acquaintance (aside from a playfully abstract mathematician or two) could even imagine the sum of two apples plus two apples not equaling four apples.

In the above discussion, as elsewhere in this book, I have endeavored to avoid the impression that a genuine Life Era has been or even necessarily ever could be fully established and maintained. Instead, I have used words such as "development," "growth," and "evolution," implying that we and perhaps other equally or more advanced life forms are steadily progressing in life's quest to best matter. Elaborating upon a convenient (and as yet hypothetical) division of galactic technological societies originally proposed by the Russian astrophysicist Nikolai Kardashev and amended by the American theoretical physicist Freeman Dyson, I suggest the following categories among life's cosmic expertise.

A type-I civilization is capable of controlling the resources of its parent planet; on Earth we shall soon approximate this level of

. . . categories among life's cosmic expertise.

technological competence, for as we begin to master global weather and more positively influence our ecology, we should attain type-I status within a few generations. A second step in the development of a Life Era, a type-II civilization, entails the manipulation of one's parent star system; as our terrestrial energy resources necessarily become depleted over the course of the next several centuries, our industrial base will be forced to adopt a solar economy wherein we shall learn, if not to control, at least to utilize more effectively the vast quantities of energy constantly released by our Sun. (That might even include restructuring our Solar System to create an artificial biosphere around the Sun, a project that our descendants could realistically accomplish within a few millennia, if we assume a modest economic growth rate of one percent per year.) In turn, a type-III civilization would be capable of exploiting the energy and material stores of an entire galaxy; such a truly advanced community would be tantamount to a galactic empire, and it is entirely unclear if such a status could be fully achieved. Even so, some researchers have estimated that once an advanced civilization begins leaving its parent planet, life will biologically "radiate" into every galactic niche within several million years. Of course, a million years is appallingly long compared to a human lifetime, yet it's likely to be a relatively brief episode in the history of any long-lived galactic society. Galactic colonization of this sort would not require space travel to exceed light velocity; indeed, such voyages need not surpass even one percent of this ultimate speed limit—a velocity currently attainable with our (rather primitive and yet unapplied) knowledge of nuclear spacecraft propulsion. Nor, therefore, is a type-III civilization impossible. Rather, a sufficiently motivated society enjoying industrial growth at current (Earth-based) exponential rates could probably progress from type-I to type-III status within several million years.

Incidentally, the time scale for galactic colonization just mentioned is often used to argue that we must be alone in the Milky Way, for if other intelligences have emerged in the past, they might be expected to have already colonized virtually all habitable planets in our Galaxy. The fact that we don't see evidence of their colonization—for example, interstellar radio communications, settlements on nearby planets or moons, unworldly stellar engineering projects, even evidence that *we* are the colonists of another galactic civilization (science fiction and charlatanistic claims notwithstanding)—implies that no one else is out there. Alas, numerous pitfalls plague this sort of argument. Advanced life forms might consciously decide to limit or even to terminate their technological growth, either out of necessity or ethical scruples, before attaining type-III status. Or they might simply lose interest in pursuing grander objectives, in effect reach the stage of physical and mental stagnation of which I spoke in *Cosmic Dawn*. A good friend of mine, radio astronomer John Ball, even hypothesized that our Milky Way might well be already ruled by a type-III civilization, in fact governed so masterfully that its inhabitants have placed aside our sector of the Galaxy, as we do on Earth with wildlife preserves, so that they can learn more about cosmic (especially cultural) evolution by studying us in our natural habitat—a proposal suggesting that we effectively live in a galactic zoo! As for myself, I think it reasonable that we are among the earliest of technologically oriented species to originate in the Galaxy, though again this results from evolutionary fortune, not from any anthropocentrism. As noted in the previous chapter, the development of organized structures, following the rules of thermodynamics and information theory, depends on the expansion of the Universe, for it is this grandest of all cosmic changes that drives a nonequilibrium between the radiation and matter temperatures. As generally implied there, the onset of life and intelligence requires certain minimum (though broadly specified) times for their origin, and thus technological intelligences might well arise and evolve more or less in parallel throughout the Universe. If true, then our Galaxy is potentially populated with myriad civilizations at or nearing type-I status, perhaps even a fraction of them as vanguard societies enjoying type-II status, but none deserving to be labeled "type III." All such extant civilizations would then be surprisingly close to us in their evolutionary progress. This is my reply to Enrico Fermi's famous query—namely, if extraterrestrials densely populate our Galaxy, then "Where are they?" The answer is that the phenomenon of technological intelligence might be only now "coming on line" in the cosmos, for the conditions needed for the emergence of such intelligence have themselves only recently emerged.

Each of these three types of cosmic civilization is but a step in the shaping of a Life Era. To achieve a genuine Life Era akin to the Energy and Matter eras that precede it would require life's control of resources on a scale at least as large as the galaxy clusters that top the hierarchy of cosmic matter and perhaps on a truly astronomical scale comparable to the entire Universe. In effect, though types I–III are necessary levels in the progressive evolution of technologically intelligent life forms and therefore in the *development* of a Life Era, the mature *establishment* of such an era would require the maintenance of a type-IV civilization. Here a "universal civilization" would presumably be the pinnacle of life's evolution in the cosmos and would permit intelligence to effectively divorce itself from matter, dominating it at will.

Is a type-IV civilization attainable? Could life evolve to the point of completely overwhelming energy and matter on all scales, even to the extreme that the environment under control exceeds the dimensions of a typical galaxy cluster? No one can realistically answer these questions, though I do acknowledge some colleagues who glibly respond, "No, that's ridiculous." I too think it unlikely that life, regardless how advanced, could ever dominate the entire Universe, for the implication here is that the intellectual prowess of life itself, rather than the density of matter, is the key factor determining the ultimate destiny of the Universe.

Nonetheless, I often wonder if we, magically present during the initial Energy Era, would have judged matter capable of ever overwhelming the radiation of the early Universe. Recall from Chapter 4 that although the cosmic expansion during the first few minutes had caused the temperature and density to diminish enough to begin the formation of atoms, the completion of the phase transition from plasma to the neutral state took time. The establishment of the Matter Era was not instantaneous; its development required upwards of a million years or so. And from a microscopic "viewpoint," which among the first few neutral atoms emerging from the energy-laden fireball would have "thought" it possible that atoms everywhere would eventually and collectively exert a strong enough influence to dominate energy on a truly grand scale (indeed, on the *grandest* scale if the Universe be closed)? Likewise, who among us can now imagine, from our rather parochial perspective at the dawn of the Life Era, what might become of technologically oriented organisms at the broadest and most advanced realm physically and biologically attainable?

Clearly, then, we are only now entering the transitional phase

193

separating the dominance by matter from the dominance by life; we are to the Life Era as the first atoms shortly after creation were to the Matter Era. But if civilizations do require "merely" a few million years to evolve from type-I to type-III status, then who can argue convincingly that type-IV status is impossible, given the astronomical time scales available in the future? And even if the great majority of civilizations fail to get their acts together sufficiently to survive beyond even type-I status, *some* civilizations might well succeed. Indeed, the steady progression toward the establishment of a mature Life Era requires that only one such civilization persist for a great duration. Furthermore, once any civilization evolves beyond type-I status, then its chances for long-term survival are considerably bettered if only because the dispersal of some members of its species beyond its parent planet will render that civilization virtually invulnerable to terrestrial catastrophe.

These were among the origins of my comment at the end of *Cosmic Dawn* that "the destiny of the Universe may well be determined not only by matter but also by the life that arises from it," a speculation that I have since found others made before. For example, some ten years prior though without elaboration, Dyson stated: "Life may succeed against all the odds in molding the Universe to its own purposes. And the design of the inanimate Universe may not be as detached from the potentialities of life and intelligence as scientists of the 20th century have tended to suppose."

As ethereal as these issues of galactic civilizations and life control may seem, I regard little of this discussion to have been metaphysical. Admittedly we cannot currently bring the scientific method to bear on the validity of these concepts by experimentally testing them. This is true because the establishment of different types of civilizations depends more on the legislated laws and decision-making prowess of intelligent life than on the natural laws of physics. But one thing is certain: Successful transitions from one type of civilization to another and the growing participation of intelligent beings in the evolution of the Life Era assume survival at each and every step of the way.

Specifically, for us on Earth, the presumption is that we shall surmount perhaps the most difficult "hurdle of change" by dynamically stabilizing our planetary society during the crucial period of the next century or so. But shall we survive beyond the dawn of the Life Era? Is there some tool, institution, or attitude to help guide us along the way?

"The time has come," the Walrus said,
"To talk of many things:
Of shoes—and ships—and sealing wax—
Of cabbages—and kings—
And why the sea is boiling hot—
And whether pigs have wings."
—Lewis Carroll

Examining Earth up close—for instance, the gardens in my backyard—I find remarkable the resiliency, adaptability, and diverse sense of survivability among the many varied plant and animal life forms. By contrast, looking at our planet from afar—as in those NASA "big pictures" of Earth drifting in space—I gain an acutely different impression, for I sense the frailty of our planet's thin biospherical membrane. The ecosphere of Earth is subject internally to chemical and thermal assault as well as externally to belts from innumerable asteroids and comets. But also concealed is a peculiar threat, nontangible yet just as real, stemming from an invisible pattern of competing national sovereignties tossed like a patchwork quilt across the surface of our world. With this planetary perspective then in mind, not just the provincialism endemic to my own back-yard, I repeat the question expressed at the end of the previous section: Shall we survive as a globally intelligent yet globally fragile civilization? Not entirely facetiously, for genuine dilemmas lurk before us, perhaps I should inquire of our survivability despite our technological intelligence.

When I wrote of the future in *Cosmic Dawn*, most of my scientific colleagues expected me to address primarily and objec-tively the fate of the Universe—the usual innocuous comparison of open and closed cosmologies—or at least the destiny of our Sun some billions of years hence. Instead, I confined my discussion to subjective local issues; local in time, over the next few generations, and local in space—namely, our civilization on planet Earth. For I believed then, and reaffirm now, that human actions during the next few decades to a century will largely determine if our species is to have a future. Many of the problems now threatening our species' longevity—from the foremost issues of overpopulation and nuclear warfare to genetic degeneration and environmental pollution among a host of other ills—are unlike any confronted by our ancestors. Ours are *global* problems. For the first time in history civilization faces not just village, national, or even regionally international problems but large-scale debilitative issues that, should they go unchecked, threaten not merely to lessen the quality of human life but, without exaggeration, to extinguish it over much of our planet. And it is quite shortsighted to argue, as I have heard stated glibly many times, that humanity will prevail forevermore because it always has managed to survive in the past despite a steady stream of political, military, economic, and natural crises over the ages. The greatest problems we now face, alas, those fated to challenge inhabitants of Earth from this day forward, are of a scope and content the likes of which have never before been encountered on our planet.

Here is a succinct statement, both to touch base with earlier thermodynamic arguments and to restate the crux of our plight from an evolutionary viewpoint: As the process of energy flowing through material structures evolves—in fact, to the extent now shared by technological creatures on Earth—life forms eventually become sufficiently advanced to determine how and when changes occur. Human actions become just as much a part of evolution as the impersonal movements of atoms or galaxies. So how and when *should* we change? Or should we change at all? Clearly, answers to such deceptively simple questions are partly subjective, and thus, I feel obliged to acknowledge that henceforth in this book my objectivity is necessarily dulled. Even so, I subscribe to the admonition once offered by the critic George Steiner—to wit, "To ask larger questions is to risk getting things wrong. Not to ask them at all is to constrain the life of understanding."

Of all the implications for the Life Era, the most important in my view is that we, as the dominant species on Earth, must develop—evolve, if you will, and quickly, too—a global culture. We need to identify and embrace a form of planetary ethics that will guide our attitude and behavior toward what is best for *all* humankind. In short, humans have got to acknowledge that we are first and foremost citizens of a planet, only secondarily members of nationalistic countries with ever-changing boundaries. It is essential that we broaden our outlook in all respects.

Ethics. My dictionary asserts, among other definitions, that ethics means "conduct recognized in respect to a particular class of human actions or a particular group, culture, etc." Formerly the nearly exclusive purview of philosophy and religion, a viable ethic for today's world is in my view no longer provided by either of these venerable institutions. Lest I be misunderstood, in the next few paragraphs I shall attempt to clarify my criticisms of philosophy and religion *as a source of modern ethics.* And lest someone think my panacea is science, I shall include science in my brief polemic, for it too, alone, will not likely provide the ethics I feel we need to seek.

Recognize that my concern here is worldly, Earthly. Whereas ethical values have, for the most part, been historically limited in scope (like unquestioned loyalty to some tribe) or even regionally widespread and more sophisticated (like those introduced into human affairs by Christ), today's set of ethics, like the global problems they must check, need to be of a more planetary, even universal nature. We must redefine ethics to denote "conduct *collectively* recognized with respect to *all* classes of human actions comprising our *global* culture" (my amended definition), and we must strive to make those ethics a practical reality by simultaneously casting them both broadly enough to apply to *Homo sapiens* in toto and flexibly enough to incorporate the process of change itself. Appropriately, the heart of the required ethics is change and adaptability, not a set of rigid, immutable rules.

Consider philosophy. Once the symbol and guardian of ethics in human society, philosophy has in my view forfeited the influential position it held throughout much of human endeavor. Mostly dated, abstract, anthropocentric, and astonishingly specialized, traditional philosophy has seemingly lost its compass in providing aims and objectives for human society, which in our day and age is both multinational and technological. Regrettably the great synthesizer is virtually extinct, the legacy and philosophy *of approach* left by Socrates, Plato, and Aristotle, among others, rather thoroughly squandered. Even if we could look to philosophy for worldly guidance, which of the many competing systems of thought should we espouse to the exclusion of all others: rationalism or spiritualism, existentialism or élan vital, or even a revival of essentialism, among legions of other philosophies proposed throughout history? In a related vein, which politico-economic doctrine might be most compatible with a global ethics: capitalism, communism, or perhaps a return to theocratic rule, likewise among numerous governmental systems? Though we can and should fully recognize the contributions made by many of history's elder intellects (as I tried to do regarding the concept of change in Chapter 2), the worldly adoption of one or another of their sectarian principles would immediately incur the wrath of large and influential groups as well as the noncooperation of many nations. The very fact that there exist so many different philosophies testifies to the unverifiability of their conclusions, however reasonable, and therefore to the unlikelihood of their ever becoming universally acceptable.

As for religion, I deplore its fragmentation. Though we surely live in a pluralistic society, what is one to make of the fact that our civilization is ministered to by some ten thousand different faiths, each with its own set of beliefs, dogmas, and often (yea, sometimes violent) insistence that its is the "one true faith"? How can the institution of religion, given its surprising lack of worldly cohesiveness, guide today's society toward what will surely need to be a coherent framework of understanding for the good of all peoples? And while I certainly welcome diversity and pluralism among humanity, how can we possibly base a planetary outlook regarding any principle, let alone one as subtle as ethics, on even a major theology, whether Buddhism, Catholicism, Hinduism, Islam, Judaism, Protestantism, or Unitarianism? As with the dilemmas just expressed regarding choices of philosophical and ideological systems, I perceive no way to decide which system of beliefs could realistically become the effectively official global religion without inviting active hostility among the thousands of competing faiths not so chosen. Note that I am not claiming such a choice to be difficult, rather that it is unproductive, considering religious proclivity to grant dogma precedent over reason. Who among today's ecclesiastics takes the larger view, addressing the present and future, not just the past, while advocating unification that might provide a holistic sense of global well-being? Who among them speaks for planet Earth, as materialistic as that may sound?

Nor will science alone (and even less likely in conjunction with its practical by-product, technology) provide the kind of ethics required to attain the Life Era. Here I mean broader *societal* ethics, not the highly regarded and remarkable scientific ethics that keep fraudulent science to an absolute minimum. Despite the moral concerns on the part of some scientists and professional societies— witness the 1975 Asilomar conference of biologists questioning the proprieties of genetic engineering, and the moral distress expressed by many physicists associated with the development of the atomic bomb—the great majority of my colleagues are unaware, or at most mindful but inclined toward benign neglect, of the socioethical implications of their work. (Even at Asilomar the debate concerned public health consequences, not the larger ethical and moral issues, of research in recombinant DNA technology.) Though we seldom admit it, our excessive specialization makes us astonishingly myopic, blinding us to the wider cultural impact of our research—at least while working at the height of our careers. Later in life and often in retirement, when scientists are usually dismissed as "no longer active" or even potentially senile, many of science's most eminent scholars begin examining the broader consequences of their work. Regrettably, often only when their influence has eroded by a sort of career entropy do they discover that some of their earlier research carried global implications. Is it possible that the duties and responsibilities, normally coupled one for one in the legal profession with every right and privilege, have not comparably grown in the last many decades with the rapid expansion in basic scientific research? Should we not awaken our attention to a formal code of socioethics among scientists as a sort of quid pro quo for the right to freedom of research?

All this is by way of encapsulating how in my view neither philosophy nor religion *alone*, alas, not science alone either, is likely able to generate a compelling set of global ethics required to aid humankind at our current turning point. Granted, each of these institutions might *think* they do, but my claim is that none of them individually can be counted on to provide an ethical standard needed for the human species to endure rapid, global, often self-induced changes in our politico-economic and especially technological environments. The twentieth-century philosophical writer Will Durant well articulated our growing predicament: "We suffocate with

199

uncoordinated facts; our minds are overwhelmed with sciences breeding and multiplying into specialistic chaos for want of synthetic thought and a unifying philosophy."

If not to one of these established institutions, then where do we turn for guidance, for survivability, at least for a sense of hope? The answer in my view is that we look to an amalgam of these three (which is, among other reasons, why I declared at the outset of the Prologue that we are entering an age of synthesis), provided we can identify a common denominator or underlying unity to which each of these three institutions can subscribe. Fortunately we do know of a unifying pattern pervading all; that common basis is evolution. Affecting everything in the Universe, from galaxies to snowflakes, from stars and planets to every aspect of life itself, evolution—developmental change—pertains to all objects, societies, civilizations, and institutions. In particular, the concept of evolution, invented by philosophy and now fully embraced by science, is acceptable to all but the most fundamentalist religions. Its broad approval is why an appreciation and understanding of evolution in its most awesome sense—cosmic evolution, a scientific philosophy capable of applying the tools of technology to the time-honored questions first posed by philosophers and theologians—can provide a map for the future of humanity.

For those who would promptly balk at this proposed synthesis, seeking, instead, to preserve the status quo by resorting to traditional institutions, let me say this. In my mind, all philosophy and religion seek a static truth—a one true dogma on which everyone can converge. But modern science has now (re)discovered such a fixed reality and bolstered it with observational evidence; it is the process of change itself. In an intriguing apposition of terms befitting the age-old ideas of Heraclitus, change has a genuinely static presence in the Universe. What's more, in the new nonequilibrium thermodynamics, change is the root of all organized stability. Once we have adjusted our thinking to accept this permanence of change, we can proceed, if need be, to change that change in ways that lead to beneficial evolution rather than devolution, entropy, and extinction. Of great import, the process of change and the "big thinking" that cosmic evolution represents can form the essence of an intellectual vehicle needed to develop, indeed to evolve, a worldly set of ethics.

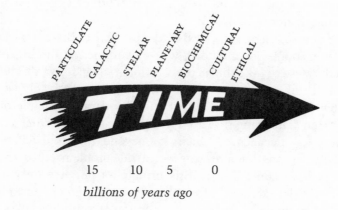

15   10   5   0

billions of years ago

. . . we must make synonymous the words "future" and "ethical" . . .

In *Cosmic Dawn* I labeled the seventh great construction phase of the Universe "future evolution." I now tell my students that if our species is to survive to enjoy a future, then we must make synonymous the words "future" and "ethical," thus terming our next grand evolutionary epoch "ethical evolution."

The Universe does conform, not to a grand design but to the chancy dictates of evolution, including, presumably, the developmental advances required for technologically intelligent life forms to survive. After all, since we have recently become the agents of change on Earth, we must now begin playing an *active* role in the process of evolution. And I maintain that this active role must begin with a collectively recognized set of ethics or principles suited to the preservation of all humankind. Furthermore, like the evolutionary changes that in turn originated and developed particles, galaxies, stars, planets, biochemicals, lives, and cultures, transition toward the next step of globally conscious life forms is a universal phenomenon. All technological beings, on any planet, must evolve a planetary ethic, lest they be unprepared to endure the by-products of technoculture. In fact, implicit within our cosmic evolutionary paradigm is a transcendence of the Darwinian principle of natural selection, a loftier standard that I call the *principle of cosmic selection*: Those technological civilizations (of any type on any planet)

201

that recognize the need for, develop in time, and fully embrace a global (even a galactic and then a cosmic) ethics will survive, and those that do not will not.

Of course, there always exists the possibility that no species on any planet, ourselves included, will be sufficiently intelligent and especially wise to take the next evolutionary leap forward. Though I prefer to think otherwise, the Universe could conceivably be regulated by a natural (or even supernatural) "cosmic principle of self-destruction," dictating that all development abruptly stops roughly within a few decades to a century beyond the time when each civilization begins encountering worldwide problems; if so, then we on Earth have come within this principle's purview only during roughly the last decade. More than just a statement of ordinary biological extinction (for here destruction is self-induced), such a principle could naturally derive from a drive toward complexity that effectively runs out of control. The rate of change might itself change so rapidly that not even technologically intelligent life could keep pace with its accelerating onslaught of global troubles, the result being that eventually all civilizations commit the ultimate devolutionary change: termination. Less of an anthropocentric statement than initially it might seem, this supposed principle of destruction would ostensibly apply to every planet, thus alleging that no one progresses much beyond our level of expertise. According to those who subscribe to it (strangely enough, mostly biologists, aside from the habitually negative sociologists), the Universe remains matter-dominated everywhere and forevermore, making no appreciable advance beyond the dawn of the Life Era.

By contrast, there are those, myself included, who prefer to opine that some civilizations—though not necessarily ours on Earth—could become smart enough quickly enough to welcome the needed ethics sufficiently to persist beyond, let us say, type-I status. Though I know nothing of the sociology of galactic aliens, my thesis here is that the way *for us* to wisen rapidly is to adopt cosmic evolution as the guiding paradigm and nouveau scientific philosophy for our time. Mine is a positive view, a synthesizing posture, and, I judge, a realistic attitude despite the onslaught of apocalyptic issues now confronting us on Earth—a vision that decidedly rejects the cosmic principle of self-destruction noted above, to be sure, one that offers a more confident, enduring, or at least optimistic prognosis. To employ cosmic evolution as an intellectual as well as a practical guide toward the Life Era is to think in dynamic rather than static terms, to forge a link between natural science and human history, to realize the evolutionary roots of human values, to renew a sense of hope.

> *Come now, and let us reason together. . . .*
> —*Isaiah, 1:18*

> *Whatsoever may be the plays on words*
> *and the acrobatics of logic, to understand is, above all, to unify.*
> —*Albert Camus*

Upon my moving to Philadelphia for a few years several years ago, the much-admired Ethical Society was one of the first local groups to kindly ask me to speak. I accepted with keen anticipation that here I would gain some insight from a cluster of thoughtful citizens who regularly discuss and debate ethical ideas and ideals. I did learn much that Sunday morning, but I came away a little disappointed.

Sitting on the platform listening to a recital of Rachmaninoff's *Humaresque* while waiting to be introduced, I glanced at the day's program. On the back I spied the society's prideful proclamation, a genuinely inspiring statement of principles, but the last sentence bothered me:

> Ethical Culture is a humanistic, religious, and educational movement inspired by the ideal that the supreme aim of human life is working to create a more humane society. Our faith is in the capacity and responsibility of human beings to act in their personal relationships and in the larger community to help create a better world. Our commitment is to the worth and dignity of the individual, and to treating each human being so as to bring out the best in her or him.

While I can understand sentiments highlighting the primacy of the individual, especially in an institution located a mere few blocks from where the framers of the American Republic declared their independence, I find incompatible the prospects for long-term survival of the human species with the often conflicting desires and lobbying efforts of virtually each and every single human being. My position has evolved during the last many years to the point where I now embrace the cause of humanity as a whole, including, if necessary, the greater regulation of society and the diminished liberty of individuals that are apparently required to ensure survival of a technological civilization. To state it in terse though telling terms, people should always be free to destroy themselves but should not be free to destroy the species.

Only moments before being introduced, I thus decided to challenge the assembled ethicists by abandoning my prepared talk and supporting the cause of humanity tout ensemble even to the potential detriment of the individual, a sentiment that doubtlessly derives from my broad interpretation of "human rights" as applicable to all humanity in general at least as much as to singular human beings in particular. Later, while we sipped tea, it became clear that the clientele was skeptical of my argument. And it wasn't the first time that an especially irate opinionist had labeled me (quite incorrectly) a Marxist cosmologist, for years prior several Harvard pre-business students (laissez-faire capitalists all) had startled me with that preposterous accusation. I begged leave of the Ethical Society gathering when several elderly members began worrying about their Social Security benefits, an attitude I found most ironic given that they wished to guard the rights of the individual yet fully expected to be the beneficiary of the social state in retirement—a harsh indictment, to be sure, but one that penetrates to the heart of the issue.

The cosmic evolutionist's commitment is to life on Earth as a matter of course, an intellectual posture that admittedly implies planetism, globalism, or just plain big thinking. Unlikely at first encounter to be popular in the West with our immoderate individualism (the "me generation" overdosed), such globalism does not admit unbridled free enterprise with its resolute insistence on monetary values and personal possessions. Nor, I hasten to add, does globalism imply in any way whatever Soviet-style communism with its psychotic and repulsive totalitarian society. In the spirit of Hegel's philosophy of the synthesis of opposites, I envision as necessary for the longevity of our species—indeed, perhaps of life on Earth—at least a rapprochement, and perhaps a merger, of the individualism characterizing the world's democracies and the socialism typifying the many varied forms of communism spread across much of the rest of our planet.

Rather than employ any of the hackneyed terms previously associated with world governance, I prefer to label my Hegelian synthesis "evolutionary humanism." I do so because, to my mind, the survival of all humanity intricately involves the adoption of evolutionism as both theoretical precept and practical guide for our day and age. Such a stance is hardly revolutionary since, as suggested earlier, the three great systems of intellectual inquiry—philosophy, religion, and science—do each now seem to accept the common-denominator notion of evolution in a general sense; cosmic evolution does have merit in, and implications for, each of these grand institutions.

Furthermore, the world's foremost political and even politico-economic systems also seem to be converging as part of the evolutionary process, and rather rapidly by cosmic standards. In particular, the Soviet Union is sluggishly evolving toward a system of expanded free enterprise, all the while the United States gradually evolves toward a system of increased social benefits (what with Social Security, the welfare system, federal income taxes, the inevitability of socialized medicine, etc.). Thinking broadly, I presume nothing odd or inappropriate about this developmental change; instead, the emerging global society should be proper and opportune, and natural as well, for geopolitics too is an essential feature of the early stages of ethical evolution. Indeed, the onset of globalism is destined to occur in some form whether we intend it or not. Increased speed of transportation, nearly instantaneous worldwide communications, international economic competition, arms control negotiations and treaties, environmental anxiety concerning acid rain and Chernobyl-like incidents, socialization of capitalistic democracies, deregimentation of communistic states . . . all suggest a natural congregation toward a planetary society. The world is being made whole, and it is good.

Historical trends in political organizations portend additional hope for a world community. Paralleling the increased technical proficiency of men and women, and the consequent "shrinkage" of our planet, is the broadening scope of social grouping; early on the australopithecine society was probably dominated by family groups, after which evolved clusters of families, Greek city-states, and so on—a communal advance resulting in the creation of the relatively recent concept of the nation-state. Is it not reasonable to expect this cultural evolutionary change to continue to the level of terrestrial globalism and perhaps beyond? Should we, if we could, accelerate this culturation, thereby passing as quickly as possible through the currently adverse social stage dominated by issues of national sovereignty?

Problematic to many of us in the West (though first proposed in modern times by a U.S. president, Woodrow Wilson, and currently a prime aspiration of the United Nations, at least in principle), globalism will likely engender a dose of societal austerity, a set of legal and social guidelines urged on humanity by the "planetary society." Yet, considering the host of worldly predicaments we and our descendants on Earth are destined to confront forevermore, restriction and limitation are a necessary, and hopefully sufficient, prerequisite for survival. Let's face it, to arrest population growth, we must physically restrain ourselves; to diminish the threat of nuclear holocaust, our leaders must exert political constraint; to avert, suppress, alleviate, or otherwise to avoid any of our impending worldwide troubles, individuals must be willing to sacrifice—and that may well include curbing free will to some extent.

True enough, the dominating force among humans will be neither government nor ideology but evolution. And the job of each individual as well as the species as a whole will be to embrace developmental rather than retrogressive change. But as I see it, an unavoidable implication of ethical evolution is the requirement that a supranational state, chosen by the collective will of individuals, safeguard the species henceforth.

We must begin to realize that given the free opportunity, most people will degenerate socially, most nationally sovereign countries war ceaselessly, most civilizations crumble institutionally. Virtuous behavior in and of itself is not predominantly advantageous to survival. At one and the same time our moral potential, as Darwin once argued and T. H. Huxley repeatedly stressed, distinguishes us from all other animals, while our culture is but a mask, as both Nietzsche and Freud maintained, disguising humankind's raw animality. In no way am I belittling the acknowledged element of cooperation and mutual aid among humankind in favor of a perverse element of cutthroat competition. But it is my contention that in mind and behavior as well as in body, the human animal bears the indelible stamp of its primitive origin among the early mammals. Appeals to individuals to act properly—"in a civilized manner"—would in the long term fall on deaf ears without a system of laws and penalties enacted by our pluralistic society either governmentally or religiously. After all, civilizations are also affected by entropy, society itself by the second law of thermodynamics. To maintain civilization at a reasonably acceptable level of humane peace and order, we need to work at it, to put energy into the system, to establish and abide by a set of societal norms that help avert social chaos, anarchy, and barbarism. Such might previously have been called social or cultural evolution, but I prefer to employ the term "ethical evolution," for we must now behave more than just socially; we must act ethically and precisely in whatever ways are needed to ensure the longevity of intelligent life on Earth.

Mine is not a pessimistic attitude (à la Rousseau—nature is good, civilization bad); rather, it's a realistic one. I feel quite comfortable with the quote that Freeman Dyson, in his book *Weapons and Hope*, attributes to the father of the computer age, John von Neumann: "It is just as foolish to complain that people are selfish and treacherous as it is to complain that the magnetic field does not increase unless the electric field has a curl. Both are laws of nature." We need not despair or give up on humanity. But we must be willing to accept change, to welcome it. For change too is a law of nature. More than that, change is probably the only real ethic.

Let me emphasize that the trend toward a global society is an entirely normal evolutionary process, though admittedly one in which we, also part of nature, have become the purveyors of change. Now beginning to compete with matter at the dawn of the Life Era and exerting increasing influence over our own destiny, humanity must work holistically to develop a program of ethical evolution as well as to guide it and maintain it. To be sure, one of the strongest implications of the Life Era concept is that life, to survive let alone to shape its future, must struggle against matter and its thermodynamic dictates, much as T. H. Huxley optimistically expressed in his Romanes Lecture at Oxford in 1893: "Let us understand, once for all, that the ethical progress of society depends, not on imitating the cosmic process, still less running away from it, but in combating it."

The greatest problem before us is one of timing. As noted at the start of this chapter, it is the changing rate of change itself that is both significant and tricky. In view of the unquantifiable social aspects of the coming politico-economic synthesis, it is virtually impossible to predict the rate of convergence. Too rapid change can harm our intended synthesis, as exemplified recently by the halting of both Afghanistan's thrust toward democracy and Grenada's flirtation with socialism. I cannot stress enough that here, as elsewhere throughout all aspects of cosmic evolution, revolution is not likely to be a substitute for evolution. Still, can we afford to wait for the evolutionary process to take its natural course? Can we run the risk that even a single global issue—say, overpopulation or nuclear warfare—could remain unsolved long enough to overwhelm us? Or if the natural course of evolution proves too gradual for our troubled world, in what ways should we actively attempt to tamper with evolution by inducing change, thereby chancing a false move in today's complex geopolitical arena? Even more perplexing, is such an attempt to accelerate the process of ethical evolution an integral, perhaps necessary part of ethical evolution itself? Which of these paths *is* the natural course favored by the principle of cosmic selection?

> We appeal, as human beings, to human beings:
> remember your humanity and forget the rest.
> If you can do so, the way lies open to a new
> Paradise; if you cannot, there lies before you
> the risk of universal death.
>
> —Bertrand Russell

The safest tack might well have been to end this volume here and now, especially in view of my earlier note that a concluding member of this trilogy will explore in some detail my proposed scenario of evolutionary humanism as a means to prevail in the Life Era. But I am reluctant to pose a litany of dilemmas and criticisms without at least outlining a couple of concrete factors, incomplete and controversial as they surely are, that would likely contribute positively to the survival of our species. Frankly, I am driven by Dante's admonition that the worst place in hell is reserved for those who are neutral on ethical issues.

In view of the unquantifiable social aspects of the coming politico-economic synthesis, it is virtually impossible to predict the rate of convergence.

My program for evolutionary humanism is relatively straight-forward in principle, for it depends upon the intellectual solidarity of humankind. However, I suspect its goals will be difficult to achieve in practice since much educational and political inertia must be overcome in the developmental stage. This difficulty will likely be especially acute in the industrialized democratic nations, if only because my twofold program calls for less specialization in academe and for more globalization of the world community. Above all, it stipulates a change not as much in our psychological makeup as in our awareness of the present human condition broadly conceived.

During the last many years numerous thoughtful students have asked me (aside from my sidereal, though I still think reasonable, argument in *Cosmic Dawn* that searching for extraterrestrials is a source of curiosity enrichment and thus a key contributor to our species' longevity) what measures can be taken of a terrestrial nature, and right now at that, to help get our Earthly abode in order. My answer at first seems weak and ineffectual but, upon further consideration, becomes viable. That solution, in part, is education. Education, information flow—we might even say "human-directed neural energy flux"—can at least illuminate the heart of our worldly predicament. And as with most problems, penetrating identification—consciousness raising—is often a sizable fraction of the solution.

But by education, I don't mean more of the same kind of education now in vogue around the world, especially in the industrialized nations; not more specialized education. The inescapable fact is that our intellectual efforts are at present grossly imbalanced against generalization and synthesis. Scholars and educators virtually everywhere, and at nearly all levels of instruction, are now devoting too much attention to advancing and exploiting highly specific discoveries, to the clear detriment, if not exclusion, of interest in the wider perspective of their work or simply in the value of beauty in everyday life. Education must be broadened in quantity and in quality—in quantity to alleviate illiteracy in parts of East Africa, Bulgaria, Appalachia, and everywhere else, and in quality as an antidote to overly specialized intellectual centers at Berkeley, Cambridge, Gorky, and the like. Educational myopia is a drag on the rest of the world and an obstacle to the progress that we desire.

Succinctly stated, my view is that the root of most *global* problems is intellectual provincialism, and especially the selfishness, narrow-mindedness, and insecurity it engenders.

Apart from a few rudimentary exceptions among some lower mammals and birds, education is first and distinctly a human activity; as a cultural process that passes on accumulated knowledge from one generation to another, it is entirely confined to humans. Education is the process by means of which individual latent potentialities can be developed, and not just in the bound sense of formal schooling and training for a growing person to earn a livelihood as a responsible member of a community. With the rapid, even exponential rise in knowledge in the last few decades, we need to recognize that learning is now a permanent and continuing process; adult and self-education programs, museum opportunities, trade books with a didactic bent, and open lecture forums should be increased in breadth as well as depth and, above all else, in quality of exposition.

Education is also a principal means to help people accept change instead of clinging to illusions of permanence. For nobody, fundamentally, likes change; as lingering Aristotelians bounded heavily by thirteenth-century religious dogma, we (at least in the West) have developed a cultural aversion to change. Perhaps our resistance runs even more deeply, much like a genetic flaw, since regardless of how promising the future may be, people almost invariably prefer what they have to what is promised but not immediately available. The fabled moral "A bird in the hand is worth two in the bush" has become nonsense now that the bird in hand is a ticking nuclear time bomb. But people will nonetheless, because of some ingrained fear of change, always find excuses for preserving what they have rather than take a flier on something new. It is as though we imagine the future (starting now) will render some sort of status quo despite our clear recognition that innumerable change has been the hallmark of evolutionary progress throughout the long past. Only education can help us change this curious (egocentric, really) doublethink and to realize the benefits (if not the necessity) of change. Thereafter we can learn at least to tolerate (if not welcome) the adjustments and adaptations needed to respond to the fluxing societal conditions on our ever-shrinking planet.

Third, education is the most obvious way whereby the *whole* of society can become conscious of its past and its potential future, can recognize that what appears "best" for the individual may in fact be worse for civilization in the long run, and can become inspired to strive for a fuller realization of its aims and aspirations. In particular, education can correct our nearsightedness, our lack of awareness that our interests, individually and as a species, are often best served by taking the long rather than the short view. And if education is to aid the globalization of society, then it must adopt an international bent—again, to educate broadly and diversely, paying heed not merely to national societies and well-informed local or regional groups but to Earth's entire community of nations and peoples.

Finally, more as an end than a means, and whether for the individual or for society, we must guard against education's confining itself to objectives that are exclusively practical and have only immediate utility. By contrast, education should incorporate activities valued for their own sake, as Julian Huxley aptly expressed in an early position paper founding the United Nations Educational, Scientific, and Cultural Organization (UNESCO), ". . . knowledge for the sake of knowing, discovery for the sake of discovering, beauty because it is beautiful, art and music and literature for their power of moving the human spirit, morality for the sake of living a good life, nobility of character because it is an end in itself."

Fortunately the evolutionary humanism I envision derives from the scenario of cosmic evolution, which in turn is based largely on natural science, and it is partly from this basis too that ethical evolution stems. I say "fortunately" because natural science is already the most international activity among humans and thus provides a chief means for raising the level of human welfare across our planet. On the unfortunate side, however, most scientists are abysmal teachers, the quality of science currently taught to the nontechnical populace being of especially low caliber. This is a major difficulty of my proposed program, for without interdisciplinary instruction of high quality and commitment, students and public alike will not likely appreciate the sense of planetary ethics and evolutionary humanism needed to ensure access to the Life Era.

Aside from what should be taught (the subject matter itself) and the associated issue of how it should be offered (the quality of teaching), yet another educational problem concerns the appropriate clientele to whom we should pitch the message. In this regard I have been distressed to encounter many academic colleagues, especially in the natural sciences, who actually take pride in their poor teaching skills. Akin to those wiseacre humanists who brag of their inability to balance checkbooks, many scientists seem to relish their ineptitude at penetrating what they often claim are the "fuzzy heads of nonscientists"—intemperate expressions, again and again, of the parochial views that have helped divide the two cultures for ages. In an article I authored a few years ago for an anniversary issue of the *Harvard Crimson* as well as in remarks I once delivered within the hallowed halls of the National Academy of Sciences, I tried to combat this unhealthy attitude by appealing to our inbred selfishness. Arguing that "one of these days, perhaps soon, the general public is going to begin knocking on the front doors of science departments," I suggested that "they will inform us scientists within, despite their realization that we are probably unlocking secrets of the Universe, that they no longer understand what we are doing and that they no longer intend to support our work." The upshot will be a steady, perhaps nearly irreversible trend toward a "scienceless society," the likes of which occurred in primitive form during the thousand-year doldrums that were the Middle Ages. By saying this, I am not suggesting that we are about to abandon technology and crawl back into the bush, but the study of science for the sake of knowledge is less sacred in these days of rising fundamentalism and federal deficits. Greeted by outrage among many fellow faculty members (though roundly applauded by students seeking better teaching) and by a more diplomatic though deafening silence in Washington, my remarks had little impact on my scientific colleagues. Our academic (as well as political) system today favors men of relentless persistence and concentrated ambition rather than Renaissance men and women of broad background and experience—a far cry from that espoused by the father of the scientific method, Francis Bacon: "I have taken all knowledge to be my province." To my dismay, not only do scientists seem unfazed that the roots of such a scienceless society may be taking shape right now, but they also appear, by their actions if not their words, to be rejecting the notion that natural scientists have an obligation, a professional responsibility, to share knowledge with the general public. Such sharing is also part of ethical evolution, and not just a minor part. Will the scientific academy adapt to the changing environment and begin articulating knowledge capably, or will it effectively face extinction as an institution useful to humankind?

Since I am often misinterpreted on the issue of education, let me state in carefully chosen English words that I would not have everyone generalize his or her scholarship or exclusively teach. Clearly we shall always require highly specialized researchers to decipher the myriad details, and they in turn, being abreast of the intellectual frontiers, should make the best teachers. What I propose here is a more reasonable balance, a willingness to recognize, to welcome, even to honor occasionally, the need for generalization *as well as* specialization, for teaching *as well as* researching, for sharing *as well as* discovering. For I do adhere to the view that an overemphasis on specialization is the primary cause of philosophy's having mortgaged its true forte of synthesis, of religious fragmentation and decay, of science's having shrunk from articulating its findings generally, and of the demise of the fine art of teaching among all disciplines. It is essential that we extend our outlook if we are to extend our future. Only in this way can we achieve our goal, educationally, of producing a technically competent, ethical, broadly based, and socially responsible populace.

All things considered, I am hardly more than echoing what Louis Agassiz had to say a century ago, but which has never before enjoyed more relevance: "The time has come when scientific truth must cease to be the property of the few, when it must be woven into the common life of the world; for we have reached the point when the results of science touch the very problem of existence."

Even more poignantly, Albert Einstein made the following plea:

> It is of great importance that the general public be given an opportunity to experience—consciously and intelligently—the efforts and results of scientific research. It is not sufficient that each result be taken up, elaborated, and applied by a few specialists in the field. Restricting the body of knowledge to a small group deadens the philosophical spirit of a people and leads to spiritual poverty.

Actually the solution to this educational issue may well lie with the museums of the world. While previously allied with the Smithsonian Institution and as a former adviser to the Boston Museum of Science, I often argued that museums should take up the slack where educational institutions fail: Museums should assume the responsibility for, indeed should champion the cause of, general education at all levels—from elementary school, through college, and on to adult education. The world's major museums are well suited (though hardly well funded) to present a broader, unifying view among a wide range of subject matter, and of course some of them do. Whether they be museums of art, of science, of history, or whatever, they must begin *to teach,* to bolster their courses that synthesize information for students and adults alike, and to do so by employing master teachers who can fire the imagination of the public at large. Science and technology museums, in particular, must move away from (or at least supplement) gee-whiz gimmickry, silicon-based gadgetry, and push-button dioramas and instead present comprehensive, interdisciplinary, and amusing programs of human knowledge. To be sure, if education is entertaining (an anathema to most academics) as well as pedagogical, then the populace will respond favorably, as most students readily attest to having learned more from a "showman" lecturer than from a pedant, for the same reason that a readable book has lasting impact when textbooks are long forgotten. To my mind, then, museums can be much more than storehouses and exhibitors of artifacts, works of art, and technical instruments; they could and should be taking the initiative by playing a much greater and more significant, even vibrant role in the active dissemination of general knowledge. And although I admit a clear bias, there is perhaps no better way to do so than to use cosmic evolution as an underlying theme. To be viable, though, a cross section of the world's museums must be supported by governments to the extent that their doors remain open without charge, even if that means a diminishment of local and national support of research universities, for museums would seem to be better positioned than the specialized institutions of higher learning to help us meaningfully achieve an integrated, global, evolutionary humanism.

The second aspect of my program for evolutionary humanism is the adoption of a planetwide society—one for which globalism, altruism, and personal restraint are naturally selected as beneficial, while nationalism, greed, and even personal autonomy are deemphasized. I want to make clear that I am wholly and firmly rejecting authoritarianism, communism, totalitarianism, and the like; rather, I am urging a kind of unitarianism as an ethical synthesis between the two leading politico-economic systems of our planet. (Such a proposition is of course hardly new, as it remotely resembles a wide spectrum of other, often sweeping syntheses suggested especially in recent historical times. One that quickly comes to mind is an earlier Teilhardian argument that a merger of Marxism and Christianity [sic] is the only antidote to the many varied problems of our contemporary world—in and of itself an intriguing modernism of much earlier attempts by scholastic clerics to reconcile the contrasting worldviews of Aristotle and Christ.) Consequently, my centrist ideology is at the same time evolutionary and humanitarian, an attitude that tempers the blunt facts of scientific analysis with the moral and aesthetic values of civilization.

Of course, one might object that such supranationalism (and its inherently radial discipline) are hardly needed, especially in light of my previous call for broader-based education. If we can use education to better instill in the international citizenry a sense of ethics, so goes the argument, then we need not adopt a restrictive globalism. Our society's admirable designation of specific parking places for disabled persons is often cited as an example of increased moral sensitivity, and it is. Numerous friends have reasoned to me that education itself can foster greater awareness of the special needs of the handicapped and thus keep our automobiles out of their designated spots. But the example breaks down when transformed to global dimensions, a case of small thinking gone awry in a larger realm. On a local scale, if an insensitive, undisabled person parks illegally in such a reserved spot, a disservice is often done to an individual or a minority of individuals. By contrast, on a planetary scale, if parking reservations for handicapped persons were to represent a global issue of some magnitude, then an inconsiderate action on the part of a small number of selfish individuals (maybe even one, and not even necessarily a terrorist as much as a naïve or uncaring individual) could precipitate global chaos by their callousness. Accordingly, I would prefer to see such globally threatening scenarios guarded closely by law enforcement officials to *ensure* that reasonable rules are not violated. In short, while broad-minded education would certainly serve to raise the consciousness of humanity, restrictive measures would be additionally needed to minimize the chances that any one individual might abuse the public trust and thereby jeopardize humanity. As has been stated many times in many ways, the human species can allow its members nearly every freedom except those that endanger its own existence.

Now, I realize that the aura of unitary globalism suggests that we among the (largely) Western democracies will have much to lose, while citizens of the totalitarian states will have "everything to gain." Our highly pragmatic culture leads most citizens to think that a moral stance which brings about hardship is erroneous. But this is hardly more than a knee-jerk reaction, once again the result of small thinking and an unjust arrogance of feeling that we in the West have everything to teach others and nothing to learn from them. By contrast, big thinking, which is an essential feature of cosmic evolution, mandates the need to *dynamically* stabilize Earth's diverse societies for the good of all peoples. I emphasize "dynamically," for I cannot advocate a true equilibrium state since such a configuration lacks structure, order, and—who knows?—curiosity. Recalling the arguments of Chapter 3, we as a society must strive to attain a steady but nonequilibrium state (like those of galaxies, stars, planets, and life forms) wherein the flux of energy is high and the entropy low. To be sure, as also discussed in that chapter, it is the low-entropy system (which we should now seek in the form of an ordered, organized civilization) that is characterized by restricted arrangements and constrained properties among its constituent parts. Why should individuals (and largely Westerners at that) have to make what seems the greater sacrifice? Because it is in the most basic interest of the human genes to enhance the survival of our species. The famous game-theory paradox of the "prisoner's dilemma" points up the problem: To the untutored mind, the prisoner's best move is to betray his buddy quickly, but both intelligence and scruples strongly imply that his long-term advantage is best served by cooperation. I like to think of this part of the evolutionary process as a kind of "enlightened self-interest"; by devoting one's self to the general welfare of society, we can actually promote our own interests, the greatest of which, at least from the viewpoint of the selfish gene, must surely be survival. And survival, arguably the strongest driving force in biological evolution, will almost certainly play a principal role in ethical evolution as well.

Not that the currently communistic nations will find this global state easy to achieve. Their leadership, like ours, will also need to surrender salient features of their political and economic systems, thus relaxing much of their societal regimentation while permitting increased freedom and personal creativity. Even so, the masses now living in Communist countries will likely welcome such changes; even their leaders should have an easier time "meeting us halfway," especially if the socialistic regimes are true to their Marxist "reconciliation of opposites," which is, after all, based on the Hegelian philosophy of thesis, antithesis, and synthesis. Furthermore, much of cosmic evolution is in keeping with the Engels-Marxist-Leninist doctrine of dialectical materialism.

Be that as it may, I cannot help noting that my call for an evolutionary humanism, and the consequent damping of individuality, jibes well with our statistical age. As explained in Chapter 3, much of modern science (especially its most basic subject, physics) consists of statistical laws derived from the study of mass phenomena. For example, the most basic entities of fundamental physics, be they elementary particles or atoms, are individually inaccessible to scientific observation. Our concern is with regularities and average properties of an entire ensemble of particles, not with differences among its individual members.

By contrast, I also acknowledge that the degree of individuation tends to increase during biological evolution, until among higher animals, notably advanced mammals, we recognize qualities akin to human personality and egotism. Further, some might argue that this process did not stop with the biological emergence of humans, as the tendency has been for individuals to gradually become more culturally differentiated, from the rigid class system of ancient Mesopotamia and Egypt, through the deliberate encouragement of individuality in classical Greece, to the narcissism of post-Renaissance times. This trend I cannot dispute, yet I emphasize that as globally technological beings we are now beyond nearly all of the realm of *biological* evolution and much of the realm of cultural evolution, just as the specific atoms in our bodies have already passed through particle, galactic, stellar, planetary, and chemical evolution. All these earlier epochs are largely behind us now as we enter the realm of ethical evolution, and it is here that the shrinkage of our world has for the first time made a famine in China or an epidemic in India an issue of moral concern to the peoples of Europe and America. Whether we realize it or not, alas, like it or not, individuation (at least in the form of egotism, selfishness, etc.) is being not eliminated but subdued. Our new proximity to one another and our growing awareness of cultural diversity, as well as our new knowledge of who we are and of our position in nature, demand it. Ethics themselves engender a kind of personal restraint, restricting our individual choices. Of particular interest recently, the current overpopulation and resulting desertification of northern Africa, and especially the subsequent famine in Ethiopia and Chad, have put into true perspective the need to act collectively—as did a virtual army of world relief agencies in addition to many of the best-known musicians who aptly captured global sentiments with their rousing lyrics "We Are the World."

Recognize that globalism need not destroy or even undermine creativity, diversity, curiosity, and many of the other cherished traits that truly make us human, even humane. In contrast with totalitarianism or even authoritarianism, wherein these human qualities might well be abolished, globalism as I see it entails an harmonious coexistence between individual humans and the planetary state—a quasi equilibrium (or, again, dynamic steady-state) that, properly crafted, would still permit substantial freedom and dignity. Take art, for instance. Every work of sculpture, every painting, poem, or novel, as well as music, drama, and a host of other creative and performing arts, are expressions of personal accomplishment. What's more, the arts can provide a focus for national pride, even serving as a legitimate and beneficent outlet for nationalism in place of the usual glorification of political power or military might. In fact, with the longevity of a global civilization that evolutionary humanism well might guarantee, creativity and curiosity could conceivably reach unimagined heights, if only owing to the availability of much future time.

Sport is another endeavor whereby nationalism can be vented harmlessly yet personal achievement can be prized and honored. The Olympic movement provides the most visible example of international sporting events, though of course in recent years these games have become so politicized that their very existence is threatened. The planetary society that I envision would perforce minimize, if not remove, the competing politics from the modern Olympiad, thereby helping transfer national rivalries from the arena of hostile military battles to one of proud and vigorous athletic competition. Holding the Olympic Games on neutral soil would be a good beginning toward this ideal, and in my view it exceeds in importance current controversies over athletes' amateur or steroid status. Sport competition among us all, in fact, brings into focus the value of a globalism that disciplines society without stifling it, much as medical evidence suggests that disciplining our bodies prolongs our lives; the widely applauded disciplining of our minds through mandatory schooling early in life is another positive case in point, and it is here that education must be carried to another plane to ensure that everyone is trained from the earliest age to welcome rather than fear change.

International organizations such as UNESCO and the World Health Organization serve yet another useful purpose in hastening our evolution toward globalism while preserving measures of personal creativity and constructive nationalism. Despite the recent withdrawal of the United States from UNESCO (a hopefully temporary squabble regarding management procedures), this world commission enjoys perhaps the most success among all the United Nations' many varied organizations.

Furthermore and finally, whatever our differences may be, everyone is a participant in the process of change extending from the individual to the global scale. We are not powerless in the face of this change. Challenges and opportunities for meaningful action are everywhere: the work we choose to do; the way we choose to relate to others; the learning we choose to acquire; the causes we choose to support; the conduct we bring to our daily lives. Each of us weaves a thread into the fabric of emerging globalism. The entire world depends upon our individual choices much as our individual survival depends upon the creation and maintenance of a world community. These are some of the seemingly endless choices that currently make us agents of change—choices that will help determine, in consort with the larger community, the outcome of the planetary transition upon which we are now embarked. Not least, it is by means of such conscious choices that we can transform alienation into globalism, stagnation into learning, cynicism into caring, despair into hope.

Humanity now faces the greatest challenge that any neophyte technical civilization could possibly encounter on any planet in the

Universe. We have virtually and simultaneously gained the ability to unlock secrets of the Universe as well as to destroy ourselves, ironically in each case by means of the same technology. This is the dawn of the Life Era. This is also why we need to create a scientific philosophy. With it will emerge a comprehensive understanding of who we are, where we came from, and how we fit into the cosmic scheme of things. Critically important, this new outlook will have built into it the foundations of a global culture as well as a set of ethics to help us sustain that culture. Stated in the most dynamic terms I know, cosmic evolution empowers us to think big, to recognize our planetary citizenship, to help us realize the need for more benevolent understanding while progressing along the arrow of time. Knowledge and compassion—these are the twin guides to the future of our species. To express it grandly but no less truly: Cosmic evolution is the scaffolding for a modern humanism, an evolutionary humanism.

Only by learning to appreciate our universal heritage, to embrace future-oriented cosmic evolution as our principal ethos of life, much as earlier peoples adopted past-oriented traditional philosophies and religions, can we survive. The solution need not take the form of a battle; rather, it can and should be a synthesis and reconciliation of the apparently conflicting claims of antiquity and modernity, of capitalism and socialism, of science and philosophy-religion in a single evolutionary process. This is the magnanimous change that awaits us—an awesome societal transformation that will likely precipitate adaptation or extinction, in a sense life or death. To forsake change, to reject the next great evolutionary leap forward, would be, according to the principles of the Universe that created us, most unconscionable.

EPILOGUE

Toward a
Modern
Metaphysics

There are more things in heaven and earth, Horatio,
Than are dreamt of in your philosophy.
 —William Shakespeare, Hamlet I, v

COSMIC EVOLUTION STIPULATES that complexity arises from
simplicity, or we might say, order from chaos. Observations demonstrate that an entire hierarchy of entities has emerged, in turn,
throughout the history of the Universe: energy, particles, atoms,
galaxies, stars, planets, life, and intelligence. This increase in complexity with the inexorable march of time—a distinctly temporalized Cosmic Chain of Being—violates no laws of physics. Nor
is the principle of ubiquitous change novel to our modern worldviews. What is new is how nonequilibrium thermodynamics now
helps us mold a holistic cosmology wherein life plays an integral
role. And how the resulting scenario of cosmic evolution can now be
recognized as a source of global, even cosmic ethics needed for the
indefinite survival of technologically intelligent species.

Without knowledge of any extraterrestrials and with humanity's only recent arrival on the threshold of the next grand period of cosmic development—the Life Era—life's destiny can only be speculative. To be sure, while I judge all my statements in this work to have been decidedly nonteleological (even including extrapolation to a potential type-IV civilization), some aspects of the prospects for life's future significance do become clearly metaphysical. Having admitted that, and in the spirit of making a gesture toward the unknown, let me now offer the following premise: As sentient beings we are currently beginning to exert a weighty influence in the establishment of a "universal life" with all its attendant features, not least of which potentially include species immortality and cosmic consciousness.

> Two roads diverged in a wood, and I—
> I took the one less traveled by,
> And that has made all the difference.
> —Robert Frost

The medical essayist Lewis Thomas has commented on NASA's greatest contribution to science—namely, those spectacular "lunar Earthrise" photographs of our planet, taken from afar by the Apollo astronauts:

Viewed from the distance of the moon, the astonishing thing about the earth, catching the breath, is that it is alive. The photographs show the dry, pounded surface of the moon in the foreground, dead as an old bone. Aloft, floating free beneath the moist, gleaming membrane of bright blue sky, is the rising earth, the only exuberant thing in this part of the cosmos. If you could look long enough, you would see the swirling of the great drifts of white cloud, covering and uncovering the half-hidden masses of land. If you had been looking for a very long, geologic time, you could have seen the continents themselves in motion, drifting apart on their crustal plates, held aloft by the fire beneath. It has the organized, self-contained look of a live creature, full of information, marvelously skilled in handling the sun.

This passage, as well as any other I have come across, captures the essence of terrestrial life at the dawn of the Life Era. Energy in the form of sunlight striking a planet rich in raw materials, plants and animals utilizing huge quantities of that energy, civilized life forms collectively storing information, manipulating matter, exhaling wastes, all the while embracing yet dominating nature on nearly every local level. Our Earthly abode does indeed resemble a living creature.

Thomas further ponders our living planet in a brief, though profound, essay entitled *The Lives of a Cell*, likening Earth to a single cell:

> I have been trying to think of the earth as a kind of organism, but it is no go. I cannot think of it in this way. It is too big, too complex, with too many working parts lacking visible connections. The other night, driving through a hilly, wooded part of southern New England, I wondered about this. If not like an organism, what is it like, what is it *most* like? Then, satisfactorily for that moment, it came to me: it is *most* like a single cell.

The notion that Earth itself is but a huge, though intricate, life form dates back, like the concept of change, millennia. Anaximenes and Aristotle apparently regarded the world as living, as did in more recent times da Vinci and Spencer (at least as regards human society), among others. Even the fashionable yet controversial Gaia hypothesis noted briefly in Chapter 2 (and currently championed by Britain's James Lovelock) stipulates that the process of evolution has endowed organisms with an innate ability to keep surface conditions favorable for themselves; the biosphere itself is theorized to comprise a homeostatic feedback or cybernetic system that has prevented drastic climatic changes on Earth throughout most of our planet's history.

With the dawn of the Life Era and its associated increasing dominance of life in the Universe, we naturally wonder what implications might pertain should these ideas be extended to the cosmos generally. If Earth *does* resemble a single cell, and if there be other sites of technological intelligence in the Universe—either independently inhabited by extraterrestrials or eventually colonized by future adventurous Earthlings—then might we henceforth regard the Universe as a multicellular organism?

Since we have no unambiguous evidence for galactic aliens, let me confine my remarks here to the dispersal of our civilization toward new habitable abodes beyond planet Earth. This might be done solely by entrepreneurial spirit, missionary zeal, or sheer exploratory drive; more likely, as discussed at some length in *Cosmic Dawn,* such dispersal will probably occur as humanity realizes that to keep all our eggs in a single basket is neither intelligent nor wise. Nor, as noted in the previous chapter, might it be ethical to do so, if we are to preserve our species for an indefinite period of time.

Paraphrasing the above sentiments about Earth as a living cell, I could imagine the origin and evolution of a "universal life form" larger and more complex than a single planet-sized "cell." Viewed macroscopically—say, over galactic dimensions—scattered sites of spreading humanity would doubtlessly intercommunicate as part of a growing network of higher intelligence. Transitioning toward the Life Era, our extended civilization would resemble a multicellular organism: "metabolizing" while utilizing increasing amounts of energy, "reproducing" as more colonies are established, "adapting" to foreign environments, even "cogitating" by transferring and storing information, all the while generally "residing" on a macroscopic scale vastly larger than the biospherical life now on planet Earth. I could even visualize such a galactic ecosystem as genuinely evolving, with some sites getting their acts together and thus surviving, others if only by chance and environmental circumstances not being so favored—a truly grand-scale manifestation of that principle of cosmic selection that I broached in the previous chapter.

My point here is that whether a given planetary civilization such as ours disperses its technologically competent species to numerous extraterrestrial abodes or many such sites serve to originate alien intelligence independently, the dawn of the Life Era signals the growth of "cosmic multicellularity"—provided the various galactic locales intercommunicate, if only electromagnetically. Such an invisible network of extraterrestrial civilizations would uncannily resemble the neural architecture of the human brain: the dendritic tentacles, the lack of hard wiring, the essence of a central nervous system. Presumably the network-to-galaxy mass ratio (much as the brain-to-body mass ratio currently used to appraise Earth's life forms) would approximate the level of intelligence among galactic civilizations. More appropriately, some such ratio might reflect the degree of cosmic ethics, for just as we need to arrest national sovereignty and all its associated hazards to survive on planet Earth, so would galactic civilizations need to restrain global sovereignty to endure in the Galaxy and Universe beyond.

Am I going too far? Is this science fiction? Clearly we cannot be certain if such cosmic networks will ever come to pass, yet I merely seek to explore the implication for long-term survival, growth, and evolution toward the Life Era. Skipping the details—an easy task since no one knows them—I can envision the evolution of a "universal life" consonant with the establishment of a type-IV civilization and the onset of a genuine Life Era. I am not saying that the Universe per se now *is* intelligent, a distinctly metaphysical posture recently championed by, among legions of theologians, the British astrophysicist Fred Hoyle. Nor am I in agreement with the (apparently) teleologically inspired attitude of today's practitioners of the anthropic principle, who essentially argue that the Universe was endowed with a mind at creation. (The anthropic principle states that intelligent life, be it rare or common, could not have evolved in a physical Universe constructed even a tiny bit differently; thus, preexisting intelligence must have designed the cosmos.) What I am suggesting is that over the course of time and in a strictly evolutionary manner, the Universe could *generate* by means of its resident life forms a sense of higher intelligence, an astronomically collective intellect. Rather than claim that cosmic evolution has been teleologically guided by an intelligent or mindful Universe, quite the contrary, I sense that the phenomenon of intelligence has, is, and will continue to develop naturally via the principles of cosmic evolution.

> Upon a slight conjecture I have ventured
> on a dangerous journey, and already behold
> the foothills of new lands. Those who have
> the courage to continue the search will
> set foot upon them.
> —*Immanuel Kant*

Knowing full well that those readers tending to dismiss as nonsense such grand cosmic schemes have long since put this book aside, I should like to take this reasoning an additional step. Con-

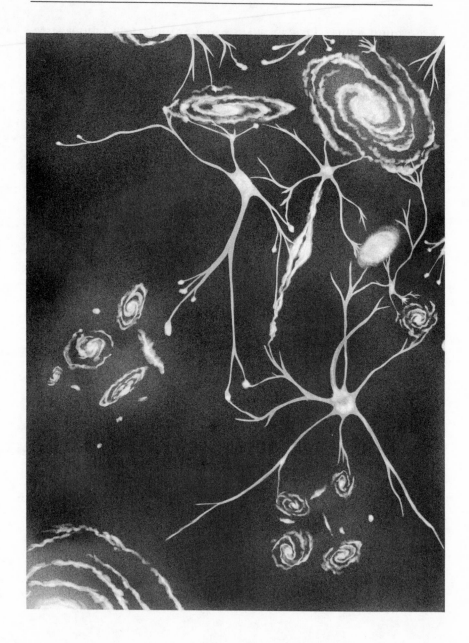

... *over the course of time and in a strictly evolutionary manner, the Universe could generate by means of its resident life forms a sense of higher intelligence* ...

228

sider immortality. By immortality, I mean unending life—a state achievable, for all practical purposes, with the onset or near establishment of the Life Era. Here I do not foresee immortality for individual life forms—you and me and our specific offspring—but rather for civilizations, perhaps for the human species or some derivative thereof, among other potentially advanced beings.

Go beyond; consider consciousness. Given humanity's current status as an exquisitely evolved agent of change, technological intelligence has gained the ability to wonder, to introspect, to abstract, to explain. Galaxies twirl, stars shine, and waves lap ocean shores, but only technologically competent life forms can begin to appreciate the grandeur of it all. Only when complex organisms arrive at the dawn of the Life Era does *the Universe* acquire self-awareness, a knowing reality. In the words of biology's Nobel laureate George Wald, "Matter has reached the point of beginning to know itself." We are, he continues, "a star's way of knowing about stars." This, for me, is life's purpose and meaning, its raison d'être—to act as an animated conduit for the Universe's self-reflection. In short, we sentient humans are now among the purveyors of cosmic consciousness. Above all else, this is what grants us, not individually but as a species, a magnanimous worth and dignity among all creatures on planet Earth, indeed, among all known structures in the Universe.

By consciousness, I mean here not merely an awareness of oneself and its external environment but also an awareness of that awareness—a rational, even technological understanding of one's relation to the Universe. This is how I distinguish organisms enjoying technical prowess from all other complex creatures nonetheless having brains. For even if a frog, a tulip, or a photo-activated garage door can sense light and some mammals may be aware of such a sensation, these nontechnical "species" are hardly able to reflect back upon the cosmic history that produced them. And if, effectively, we associate "mind" with consciousness, then my definition connotes a separation of mind and brain. To express it alternatively: While brains are a clear prerequisite for the Life Era, this same Life Era is tantamount to manifesting "mind over matter." (For example, a paramecium displaying an awareness of its environment and a chimpanzee being apparently aware of itself as well as its environment are, by the above definition, not conscious, for I have encoun-

. . . life's purpose and meaning, its raison d'être—to act as an animated conduit for the Universe's self-reflection.

tered no convincing evidence that these among all other life forms, save humans, *know* of their own awareness.) Ludwig Feuerbach, the nineteenth-century German philosopher, expressed it beautifully:

> The animal is sensible only of the beam which immediately affects life; while man perceives the ray to him physically indifferent, of the remotest star. Man alone has purely intellectual, disinterested joys and passions; the eye of man alone keeps theoretic festivals. The eye which looks into the starry heavens, which gazes at that light, alike useless and harmless, having nothing in common with earth and its necessities—this eye sees in that light its own nature, its own origin. The eye is heavenly in its nature. Hence man elevates himself above the earth only with the eye; hence theory begins with the contemplation of the heavens. The first philosophers were astronomers. It is the heavens that admonish man of his destination, and remind him that he is destined not merely to action but to contemplation.

In writing this Epilogue, I deem it in no sense egocentric to claim that we humans are the only known consciousness in the Universe; given the current lack of evidence for extraterrestrials, I am merely conforming to my (above-stated) definition. Nor—I must repeat—do I regard any of my statements as teleological, for I fail to see why consciousness must be part of any grand design. Instead, I claim, though I cannot prove my claim or even envision an experimental test for it, that consciousness is but yet another, albeit awesome, evolutionary step in the ongoing complexification of the Universe. Whether of a singular nature (should we be alone in the cosmos) or as part of a collective enterprise (among numerous, federated extraterrestrial intelligences), consciousness most likely has an evolutionary origin. Hence, I feel in no way compelled to endow the Universe at the time of creation with an ingrained consciousness that guides the evolution of all things. Rather, I judge on the basis of reason alone (hence my appeal to experimentalless metaphysics) that consciousness differs only in degree and not in kind from the many varied ordered structures that have, successively and successfully, arisen with the march of time. Thus, I extend my cosmic hierarchy: energy, particles, atoms, galaxies, stars, planets, life, intelligence, and now consciousness. Resembling Nietzsche's "higher plane of future organization," Teilhard's "Omega

Point," and even Shapley's "supra-Universe class," consciousness for me is in actuality none of these as much as simply and straightforwardly a natural consequence of the evolution of energy, matter, and life in the Universe.

If we chance to think big once more, might not the onset of universal consciousness—carbon-based or silicon-based—herald yet a fourth momentous era in the history of the cosmos? Could life's heir apparent be already within our infantile minds, concealing its most elementary potential for controlling and manipulating us by means of an ever-increasing complexity of our minds? Alas, any attempted explication of an Era of Consciousness (or, as one of my students once proffered, an Era of Moral Abstraction) best awaits another occasion.

Throughout this book-long essay I have used, largely in a scientific context, words and phrases such as creation, immortality, reincarnation, global and cosmic ethics, cosmic selection, even universal life and consciousness. As we enter the new age of synthesis, such time-honored terms, once believed outside the realm of science, are now coming within the purview of scientific thought. Indeed, these are the kinds of ideas—holistic ideas, humanistic ideas, cosmic-evolutionary ideas—that we shall need to embrace if life forms are to endure to experience the Life Era.

Children of the human race,
Offspring of our Mother Earth,
Not alone in endless space
Has our planet given birth.
Far across the cosmic skies
Countless suns in glory blaze,
And from untold planets rise
Endless canticles of praise.

Should some sign of others reach
This, our lonely planet Earth,
Differences of form and speech
Must not hide our common worth.
When at length our minds are free,
And the clouds of fear disperse,
Then at least we'll learn to be
Children of the Universe.
—John Andrew Storey

APPENDIX

A Mathematical Guide to the Three Eras of Cosmic Evolution

[for those readers familiar with the calculus]

Although matter only later emerged from the energy of the early Universe, it is pedagogically advantageous, paralleling our discussion in Chapter 4, to quantify first the role of matter and thereafter the primacy of energy. In this way the initial great change in the history of the Universe—the transformation from energy to matter—can be mathematically justified. Likewise, the second great change of cosmic evolution—that from matter to life—is quantified in the last section of this appendix.

233

Matter

The galaxies provide important cosmological information regarding the bulk nature of matter in the Universe. The key observations are embodied within (Edwin) Hubble's "Law," a discovery named after the American astronomer who first noted in the 1920s a correlation between two of the most basic properties among galaxies. (Here I have placed "law" in quotes to stress its empirical nature, for the discovery is not based on any firm physical principle.) The figure below illustrates the essence of this relationship, where the dots represent velocity and distance data observed for numerous galaxies and the dashed line comprises the best linear fit to those data.

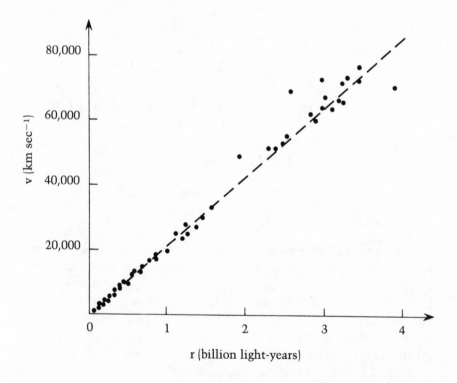

Hubble's Law expresses the observational finding that the radial velocity, v, with which a galaxy recedes is linearly proportional to its distance, r, from us. (Stated more clearly, this law accounts for the observed fact that the distances separating the galaxies, or, more precisely, the clusters of galaxies, are increasing with time.) Since r is a function of time, t, in an expanding Universe, we symbolically write this law as

$$v \propto r\{t\},$$

which, after the insertion of a proportionality constant (equal to the rate of universal recession—namely, the slope of the dashed line in the above figure), becomes

$$v = H\, r\{t\}.$$

(Here I use curly brackets to remind us that r depends explicitly on time.) On the basis of the best observations to date, H, called Hubble's constant, equals 22 ± 6 kilometers per second per million light-years, thus specifying that each galaxy recedes with an additional velocity of 22 km sec^{-1} for every additional million light-years of distance from us. (The error, $\pm$ 6 km sec^{-1} Mly^{-1}, arises from the inherent inaccuracy of the observations of some of the more distant galaxies; such objects are very faint, making their measurement and analysis a difficult task.)

Assuming (as do virtually all contemporary astronomers) that the recessional motions of the galaxies delineate the expansion of the Universe, we can use Hubble's Law to derive the time elapsed since the expansion began—namely, the age of the Universe. To do this, recall that distance traveled equals velocity times the elapsed time—that is,

$$r\{t\} = v\, t,$$

which, when substituted into Hubble's Law, reduces to

$$t = H^{-1}.$$

Using the observationally determined value of H (and properly canceling the units), we find

$$t = (15 \pm 4) \times 10^9 \text{ years.}$$

This is an estimate of the current age of the Universe, a value rounded off in this book to equal 15 billion years. (I say "estimate," for I have neglected cosmic deceleration owing to the accumulated mass of the Universe, a correction that is relatively unimportant given the main thrust of this Appendix.)

Now, leave these observational results for a moment, and consider a theoretical argument. Imagine a spherical shell of mass, m, and radius, r, moving outward with a velocity, v, from some central point, as shown in the accompanying figure. (Classically we may view this sphere as a very large, isotropic gas cloud—in fact, one larger than the extent of a typical galaxy supercluster which comprises the topmost rung in the hierarchy of material coagulations in the Universe.)

Employing a central dogma of modern physics—the conservation of energy—we can set the sum (per unit mass) of the kinetic and (gravitational) potential energies equal to a constant—namely,

$$v^2/2 \; - \; G \, M/r\{t\} = \text{constant.}$$

Here, G is the universal gravitational constant equaling 6.67×10^{-8} dynes cm^2 $gram^{-2}$, and the total mass, M, of the matter inside the sphere having density, ρ_m, is given by

$$M = 4/3 \ \pi \ r\{t\}^3 \ \rho_m.$$

(The minus sign in the middle term of the energy-conservation equation conforms to the common convention that potential energies are inherently negative.) In anticipation of later comparison with relativistic calculations, the constant term on the right-hand side is usually taken to equal $-kr\{t_c\}^2$, where k is a proportionality factor related to the total energy of the system and $r\{t_c\}$ is the *current* radius of the Universe. (The Newtonian theory used here not only is simpler but also leads to many important results that are essentially the same as those of Einstein's Relativity Theory.) Substitution of Hubble's Law and of the above expression for the total mass rearranges our energy-conservation equation—to wit,

$$H^2 \ r\{t\}^2 - 8 \ \pi \ G \ r\{t\}^2 \ \rho_m/3 = -kr\{t_c\}^2.$$

At this point in cosmological models, astrophysicists often introduce a dimensionless (but time-dependent) term called the "scaling factor," R, which relates the radius, $r\{t\}$, of the Universe at any time, t, in cosmic history to the current radius, $r\{t_c\}$ at the present time, t_c—namely,

$$r\{t\} = R \ r\{t_c\}.$$

Thus, R = 1 so that $r\{t\} = r\{t_c\}$ at the current epoch, whereas, for example, in the early Universe, $R \ll 1$. (The scaling factor, R, may be thought of as representing the average distance separating clusters of galaxies.)

Before proceeding with our quantification of matter's evolution, let me make two notes that derive from consideration of the scaling factor and that will be useful to us later in this Appendix. First, the equation for the scaling factor confirms that Hubble's constant is not really a constant; like virtually everything else in the Universe, it also changes with time. To see this, take the time derivative of the previous equation—namely,

$$dr\{t\}/dt = v = r\{t\} \ R^{-1} \ dR/dt,$$

or,

$$H = R^{-1} \ dR/dt,$$

which means that

$$H \sim t^{-1}.$$

Note also that the Doppler shift, z (which is used to derive galaxy velocities, viz., $z = v/c$ for $v \ll c$, where c symbolizes the velocity of light), is conventionally defined as the difference between a standard laboratory rest wavelength, λ_L, and the actual observed wavelength, λ_o, all normalized by λ_L; hence,

$$z = (\lambda_o - \lambda_L) / \lambda_L.$$

Local galaxies that are not receding very fast have small values of z, some quite close to zero; the most distant galaxies—the so-called quasars—are rapidly receding and thus have higher values of z, though none currently observed to exceed $z = 4$. The above equation can be recast, using the scaling factor, to become

$$z = R^{-1} - 1$$

or, as is often used in the research literature,

$$(1 + z) = R^{-1}.$$

Now, by inserting the scaling factor into the above equation for energy conservation, we arrive at one of the most basic relations of modern cosmology—namely, the (Alexander) Friedmann equation (after the Russian meteorologist who first derived it in the 1920s, though he originally did so using Relativity Theory). The result (which, again in this case, is identical whether relativity is used or not) is

$$H^2 - 8\pi G \rho_m/3 = -k R^{-2}.$$

The figure below sketches the various solutions to the Friedmann equation, illustrating that the Universe can be "open" (i.e., k negative) and therefore recede forevermore to infinity or "closed" (i.e., k positive), in which case its contents will eventually stop receding and thereafter contract to a point presumably much like that from which the Universe began. The intermediate case (i.e., $k = 0$) corresponds to a Universe precisely balanced between the open and closed models, but in fact such a model Universe would eternally expand toward infinity and never contract.

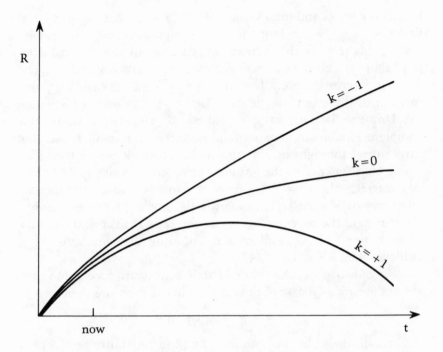

To follow the evolution of matter throughout cosmic history, we must introduce one more conservation principle. Within the sphere we require that the number of particles of matter remain constant, so that as the Universe expands the density naturally grows smaller. Taking $\rho_m\{t_c\}$ as the current matter density of the Universe, we can express this conservation principle simply,

$$\rho_m\{t\} = \rho_m\{t_c\}\, R^{-3},$$

which is just as we expect intuitively since the volume of the sphere scales as R^3. (This particle conservation equation is valid except in the very earliest moments of the Universe; many particles were themselves only materializing from primordial energy at $t \ll 1$ second, a time frame that is irrelevant to the main issue of this Appendix.)

Now, note that the Friedmann equation is clearly simplest when $k = 0$, for then it reduces to

$$\rho_m = 3\,H^2\,/\,8\,\pi\,G.$$

Evaluating for G and for a value of $H = 22$ km^{-1} sec^{-1} Mly^{-1}, we find that $\rho_m\{t\} = 10^{-29}$ gram cm^{-3} or approximately 10^{-6} atom cm^{-3}. This terribly thin spread of matter equals the "critical density" above which the Universe is closed and below which it is open.

An interesting point here is that $k = 0$ is a very good approximation to the events of the early Universe, regardless of whether the Universe is truly open or closed. To see this, examine the complete Friedmann equation and note that, for small R (i.e., for early times), the right-hand term which scales as R^{-2} will always be negligible compared to the left-hand term which scales as R^{-3}, thus guaranteeing the $k = 0$ solution as a reasonable approximation whenever $r\{t\} << r\{t_c\}$. (This point is also made clear by examining the curves of the previous figure; for the early Universe, the family of curves, $-1 < k < +1$, all overlap, meaning that any value of k suffices when $R << 1$.)

By substituting the above particle conservation equation into the simplest $k = 0$ case of the Friedmann equation, we then find

$$H^2 = 8 \pi G \rho_m\{t_c\} / 3 R^3.$$

And recalling our above note regarding the variability of Hubble's "constant," we can manipulate this equation to show explicitly how the matter density changes as a function of time throughout the history of the Universe,

$$\int dt = (8 \pi G \rho_m\{t_c\} / 3)^{-1/2} \int R^{1/2} dR.$$

Integrating and using the earlier derived relation, $(1 + z) = R^{-1}$, we find

$$t = 3 \times 10^{17} (1 + z)^{-3/2} \text{ seconds}$$

and

$$\rho_m = 10^{-29} (1 + z)^3 \text{ grams cm}^{-3}.$$

These last two expressions can be further manipulated to show the temporal dependence of the matter density—to wit,

$$\rho_m \sim 10^6 t^{-2},$$

where ρ_m is expressed in grams cm^{-3} and t in seconds.

We have therefore derived a way to quantify the evolution of the matter density throughout all time. However, hindsight suggests that it will be more useful to reexpress this quantity in terms of the equivalent *energy* density of that matter. We can do so by invoking the Einsteinian mass (m)-energy (E) relation, $E = mc^2$—that is, by multiplying the above equation for ρ_m by c^2. The figure below illustrates the change of matter's equivalent energy density, $\rho_m c^2$, throughout the history of the Universe; we shall return to it in the next section in order to compare the evolution of matter's energy density with that of radiation's energy density. (The plotted line's width or variance represents the considerable range of uncertainty in the observed value of ρ_m today.)

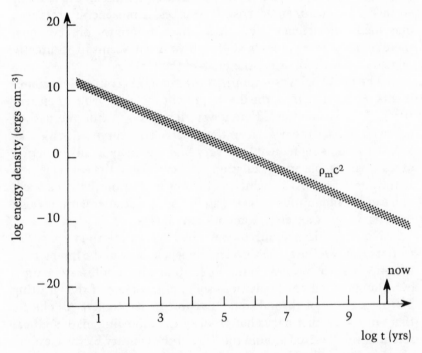

As a final point of this section, note that the most distant known material object, a quasar with the peculiar catalog name of $1208 + 1011$, has $z \simeq 3.8$, which, according to the above equations, implies that $t \simeq 1$ billion years. This means that direct study of the farthest known matter can extend our knowledge back roughly 90 percent of the past, but that it cannot help us directly penetrate any closer to creation $(t = 0)$ than about a billion years. All other known objects in the Universe are closer to us in space, and therefore their signals were launched in more recent times. Fortunately, by studying cosmic radiation (as opposed to observable matter per se), we can begin to unveil earlier epochs.

Energy

As I did in the previous section for matter, let me first introduce a relevant observation. As noted in Chapter 4, the Universe is flooded with weak radio radiation (i.e., invisible energy); regardless of the direction scanned with a radio telescope, after all the familiar signals (well-documented cosmic sources, atmospheric emission, man-made interference, etc.), are subtracted from the observations, there remains a weak hiss of static whose origin seems to be directly related to the aftermath of the birth of the Universe.

The figure below summarizes the main observational measurements of this radiation; the dots represent the observed radio intensities, B_λ, measured at different wavelengths, λ, while the dashed curve comprises the best fit to those data. This curve, which peaks at $\lambda \simeq 1$ mm, is reproducible to very high accuracy in any direction of the cosmos; observed radiation has been found to be isotropic to within one part in 10^4, which is further indication that this weak radiation is ubiquitous in space and is not associated with a single material object or even a group of such objects.

Now, the dashed curve does not have an arbitrary shape. In fact, it closely follows the wavelength dependence of the intensity of radiation emitted by any warm object. Stars, light bulbs, stove tops, even humans, all emit radiation, some of it visible (if the emitting matter is hot enough) and some invisible (when cooler). All objects that emit such thermal radiation obey the following mathematical relation, first derived around the turn of the century by the German physicist Max Planck:

$$n = 8\,\pi\,/\,\lambda^4\,(e^{[hc/kT\lambda]} - 1).$$

Here, among the as yet undefined quantities, n is the number of photons per unit volume, h is Planck's constant, k is Boltzmann's constant (not the same k as was used for the cosmological models of the previous section), T is the temperature, and e represents an

242

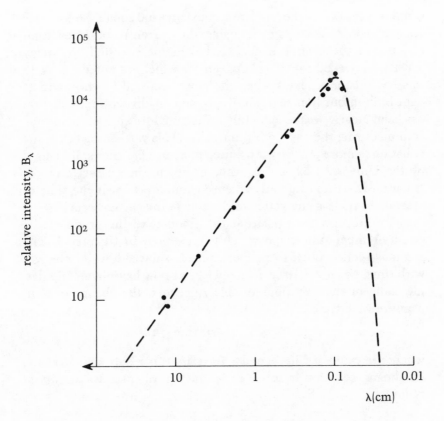

exponential function. Analysis of the spectral data (i.e., the wavelength distribution) enables us to solve this equation for the temperature, which yields T = 2.7 ± 0.1 Kelvins. This is the thermal temperature characterizing the cosmic background radiation, now greatly cooled from its fiery beginnings.

In this section we are ultimately interested in finding the density of the energy contained in the radiation so that it can be compared to the above-derived energy density of matter. We can do this by multiplying the above mathematical expression by the energy, $hc\lambda^{-1}$, of each photon and thus finding the energy density, u, of the Planckian radiation,

$$u = a\,T^4,$$

243

where a is known as the "radiation constant" and equals 7.6×10^{-15} erg cm^{-3} K^{-4}. For example, using the currently observed temperature of the cosmic background radiation, we find an energy density, u = a$(2.7)^4 \simeq 10^{-13}$ ergs cm^{-3}, which, as noted earlier in this book, equals far less than the power emitted by the smallest light bulb in our homes, again implying that this cosmic radiation has been greatly weakened while dispersing into an ever-increasing volume during the past 15 billion years. (This weakening of cosmic radiation is more properly attributed to a huge Doppler shift caused by the Universe's expansion, thus changing the character of the radiation from its originally intense gamma rays near the time of creation to its less energetic radio waves in the current epoch.)

We are now in a position to demonstrate the origin of the density-temperature graph roughly illustrated in Chapter 4. The previous section of this Appendix already showed how ρ_m changes with time (i.e., $\rho_m \sim 10^6 \, t^{-2}$); likewise, we now have, by manipulating some of the equations of this Appendix, the change of temperature with time:

$$T = 2.7 \, (1 + z) \simeq 10^{10} \, t^{-1/2},$$

where t is expressed in seconds. To verify this result, reexpress the Friedmann equation in terms of ρ_r, the equivalent mass density *of radiation:*

$$H^2 = 8 \, \pi \, G \, \rho_r\{t_c\} \, / \, 3 \, R^4.$$

Here the R^4 term derives from the fact that radiation scales not only as the volume $(\propto R^3)$ but also by one additional factor of R because radiation (unlike matter) is affected linearly by the Doppler shift. Substituting for H, integrating, and solving for $\rho_r\{t\}$, we find

$$\rho_r\{t\} = 3 \, / \, 32 \, \pi \, G \, t^2,$$

where I have also made use of the relation $\rho_r\{t\} = \rho_r\{t_c\} \, R^{-4}$. Finally, noting that the equivalent energy density of radiation, $\rho_r c^2$, equals aT^4, we conclude that

$$T = (\rho_r\{t\}c^2 \, / \, a)^{1/4},$$

which, upon evaluation, equals

$$T \simeq 10^{10} \, t^{-1/2}.$$

The figure below quantifies the earlier ρ_m–T plot of Chapter 4, and the accompanying table specifies the range of values of ρ_m and T for each of the six epochs discussed in that chapter. It is these density and temperature values, together with a sound knowledge of modern physics, that enable us to appreciate the key events and processes evident at any point in the history of the Universe.

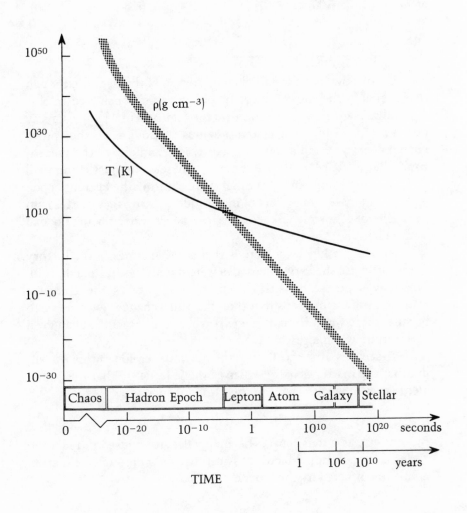

A Table of Cosmic History

Epoch	Time After Creation	ρ_m (g cm^{-3})	T(K)
chaos	$< 10^{-24}$ sec	$> 10^{50}$	$> 10^{20}$
hadron	$10^{-24} - 10^{-3}$ sec	10^{30}	10^{15}
lepton	$10^{-3} - 100$ sec	10^{10}	10^{10}
atom	100 sec $- 10^6$ yr	10^{-10}	10^4
galaxy	$10^6 - 10^9$ yr	10^{-20}	300
stellar	$> 10^9$ yr	$\sim 10^{-30}$	~ 3

Let us now make the final computation of this section, though it is arguably the most important one thus far considered in this Appendix. Here we wish to compare the "equivalent" energy density of matter, $\rho_m c^2$ (derived in the previous section), with the "pure" energy density of radiation, aT^4. Note, for example, at the present time, that $\rho_m\{t_c\}c^2 \simeq 10^{-8}$ erg cm^{-3}, whereas (the just-computed) $aT^4\{t_c\} \simeq 10^{-13}$ erg cm^{-3}. In other words, during the current epoch $\rho_m c^2 > aT^4$ by several orders of magnitude, proving that matter is in firm control of cosmic changes, despite the Universe's being flooded with radiation.

However, a key issue here is that at other times in the history of the Universe these two densities were not always in such discord. Not only do these densities themselves change as the Universe evolves, but it so happens that their ratio also changes with time. To be sure, there was a time in the past when $\rho_m c^2 = aT^4$, and an even earlier time when $\rho_m c^2 < aT^4$.

To see this, first recall from the previous section of this Appendix that the matter density scales as the volume, so that the equivalent energy density varies in the same manner,

$$\rho_m\{t\}c^2 = \rho_m\{t_c\}c^2\, R^{-3}.$$

By contrast, as noted above, the energy density housed in radiation scales as an additional factor of R because radiation (unlike matter) is linearly affected by the Doppler shift,

$$aT^4\{t\} = aT^4\{t_c\}\, R^{-4}.$$

(For these relations, as in the previous section, t represents any time after creation while t_c is the current time.)

We can now ask: When was $\rho_m\{t\}c^2 = aT^4\{t\}$? Setting the two equal, we have

$$\rho_m\{t_c\}c^2 \,/\, aT^4\{t_c\} = R^{-1},$$

or, from the previous section, since $R^{-1} = (1 + z)$, we find, upon evaluating for the (above-computed) current values of $\rho_m\{t_c\}c^2$ and $aT^4\{t_c\}$, a value of $z \simeq 10,000$. And from the equation at the end of the previous section that connects Doppler shift and time, we calculate that this z corresponds to $t \simeq 500,000$ years. Thus, some half million years after creation, the two energy densities were equal; at earlier times, $\rho_m c^2 < aT^4$.

Now, the figure below is of central import. In part, it reproduces the temporal change of $\rho_m c^2$ as was drawn at the end of the previous section. Superposed on it is the temporal change of aT^4 computed from the equations of this section. Note how the two values intersect at $t \simeq 500,000$ years, the epoch when the energy distributed between matter and radiation was last equilibrated. Relative to this crucially important turning point in cosmic history, pure energy (radiation) dominated in earlier epochs while equivalent energy (matter) ruled in later epochs.

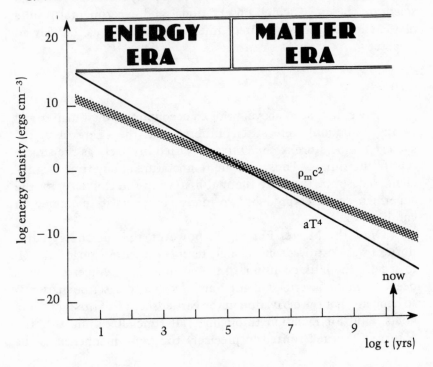

247

This crossover represents one of the two preeminent changes in cosmic history to which I have often referred in the main body of this book, especially in Chapter 4. The event, $\rho_m c^2 = aT^4$, separates the Energy Era from the Matter Era and designates that time at which the Universe became transparent; thermal equilibrium was destroyed and symmetry broke down, causing the radiative fireball and the matter gas to decouple. Photons of radiation, previously scattered many times over by the material particles (especially free electrons) of the expanding, hot, opaque plasma of the Energy Era, were no longer so affected once the electrons became bound into atoms (and eventually into galaxies and stars) of the Matter Era; the 2.7-Kelvin radiation reaching Earth today is a "relic" of this dramatic phase transition.

Thus, while studies of material objects currently prohibit us from penetrating directly any closer to creation than a billion years, analyses of the cosmic background radiation push back our direct knowledge to some half million years after the origin. This means that we can use observations of the background radiation to characterize directly the most recent 99.997 percent of cosmic history. To probe any closer to creation, we are forced to rely on theory—though *scientific* theory in which observational and experimental results play an important yet indirect role, as described qualitatively in Chapter 4.

Life

As we have seen above, prior to decoupling, when matter and energy (radiation) were still equilibrated in the Energy Era, the average temperature of the Universe varied inversely as the square root of the time, t. Thus, a single temperature at any time is sufficient to describe the early thermal history of the Universe, for the superhigh densities produced so many collisions to guarantee an equilibrium.

Once the Matter Era began, however, the gas-energy equilibrium was destroyed, and a single temperature is no longer enough to specify the bulk evolution of the cosmos. Two temperatures are needed: one to describe radiation and another to describe matter. It so happens that the derivation in the earlier section of this Appendix holds valid for radiation throughout all time. Recalling that the observed T is red shifted in precisely the same manner as is the frequency of an individual photon, and since $u \propto T^4$ and also $u \propto R^{-4}$, we now rewrite from above,

$$T_r \propto R^{-1},$$
$$\text{or} \quad T_r \simeq 10^{10} \, t^{-1/2},$$

with T_r being specifically the average "temperature of radiation" at any time, t. Matter on the other hand, now decoupled from radiation, cools much faster, and (at least for hydrogen and helium below 3,000 Kelvins in the neutral state) obeys the relation for a perfect gas. For such an ideal gas, the pressure $P = \rho_m T_m$, and also (for adiabatic [zero heat transfer] expansion) $P \propto T_m^{3/2}$. Further, since (from earlier in this Appendix) $\rho_m \propto R^{-3}$, we find $R \propto T_m^{-1/2}$, which means for t > 500,000 years after creation,

$$T_m \propto R^{-2},$$
$$\text{or} \quad T_m \simeq 6 \times 10^{16} \, t^{-1},$$

where T_m is the average "temperature of matter." (For both temperature relations, time t must be expressed in seconds.)

The graph below illustrates these diverging thermal histories as the Matter Era evolves.

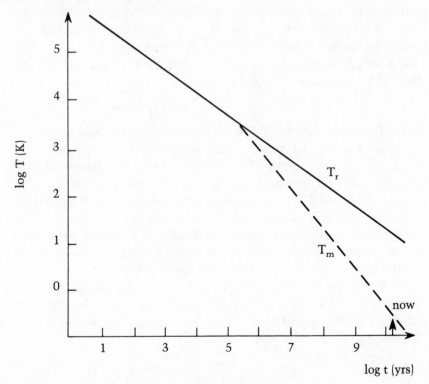

Now, as the difference between T_r and T_m grows, the cosmos, on average, departs progressively from its early equilibrium state. Hence, the Matter Era has become increasingly unequilibrated over the course of time; the expansion of the Universe guarantees it. According to our discussions elsewhere in Chapter 3, such non-equilibrium states are suitable for the emergence of order, and thus we reason that *cosmic expansion itself is the prime mover for the gradual construction of a hierarchy of structures throughout the Universe.* In what follows, I briefly show how we can quantify the evolutionary development of order in the Matter Era from the chaos of the Energy Era; the result is a whole new era—namely, the Life Era.

Recall from Chapter 3, the extent to which a system is ordered can be found by appealing to information theory; the more ordered a system, the greater the information content (or negentropy) it possesses. To be more specific, imagine a system that a priori (i.e., in the absence of any knowledge) can have a large number, B_i, of possible structural arrangements initially (hence the subscript, i). Suppose, further, that after some information is finally received, the number of possible states of the system is reduced to B_f, so that $1 \leq B_f \leq B_i$; this is true because we now have additional information about the state of the system, in particular about its structure. Quantitatively we can express these statements in terms of the net information transmitted, ΔI, which equals the difference between the initial and final informational states of the system;

$$\Delta I = I_f - I_i \propto ln \, (B_i/B_f).$$

Here I have made use of the (Claude) Shannon-(Norbert) Wiener formula that stipulates the information to depend inversely on the logarithm of the number of structural possibilities (or probability distribution) of a system; when B_f is small (after some information has been received), ΔI is large, and conversely. Note that this equation increases as B_i increases and B_f decreases, that it maximizes for a given B_i when $B_f = 1$ and vanishes when $B_i = B_f$, and that it makes information received in independent cases additive. This accords well with human intuition and experimental findings.

The above relation for information exchange has an uncanny resemblance to that describing a change in entropy, S, for a system having different numbers of microscopic states, W, before (i) and after (f) the change—to wit,

250

$$\Delta S = S_f - S_i \quad \propto \ln (W_f/W_i).$$

Assuming the proportionality constants are the same, we can substitute S_i for $\ln B_i$ and S_f for $\ln B_f$, yielding an explicit relationship between the entropy and the net information exchanged,

$$\Delta I = S_i - S_f.$$

And since $S_i = S_{max}$, because the initially large number of receivers (microstates) would relate to the maximum entropy, we find that the information gain of a system can be expressed as the difference between the system's maximum possible entropy and its actual entropy at any given time—namely,

$$I = S_{max} - S.$$

This is the origin of our statements in Chapter 3 that order, information gain $(+\Delta I)$, and negentropy $(-\Delta S)$ are intimately related.

Now, to apply these ideas to the material developed earlier in this Appendix, we use the mathematical formulation of the second law of thermodynamics. This perhaps most general law in all physics stipulates that the change in entropy varies directly as any heat exchanged, ΔQ, and indirectly as the temperature, T, at which the change occurs—namely,

$$\Delta S = \Delta Q / T.$$

Therefore, for a given amount of heat transferred (between two systems), we can reason that S_{max} will occur for minimum T, which for the general case of an evolving Universe is the system of matter characterized by T_m, as defined earlier. By contrast, the smaller, actual S will pertain to the system of radiation defined by T_r, which, after the recombination phase discussed above, always exceeds T_m. Thus, by virtue of this finite temperature difference between matter and radiation as well as the associated irreversible flow of energy from the radiation field to material objects, we obtain a quantitative measure of the information growth throughout cosmic history:

$$I = T_m^{-1} - T_r^{-1}.$$

Once again, as stressed in Chapter 4, the growth of information is guaranteed by the very expansion of the Universe (which, I repeat, is the fundamental cause of the divergence of the two temperatures).

Knowing the temporal functions for T_m and T_r (as derived elsewhere in this Appendix), we can trace the evolution of I in much the same way that the change of $\rho_m c^2$ and aT^4 were graphed earlier. The plot below sketches such an I–t diagram. Notice how the information growth was zero in the

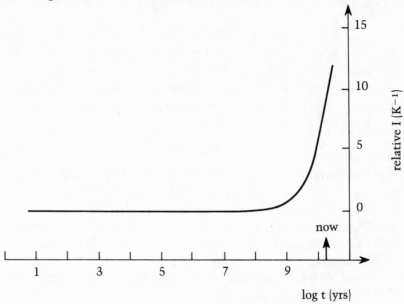

early Universe when matter and radiation were equilibrated. After recombination, however, the growth of information steadily rose as T_m and T_r diverged when the matter and radiation began departing from equilibrium; I has become substantial in recent times.

The question before us is this: Do the many varied real structures in the Universe display this sort of rather dramatic increase in order within relatively recent times? The answer is yes, and more. To see this, recall that the total energy, E, of any system equals the free energy, F, of that system plus the product of temperature, T, and entropy, S. In symbolic form,

$$E = F + TS.$$

Thus, if, for a given temperature, the entropy of some system is to be decreased—and this is the essence of order and organization, as suggested in Chapter 3—then the free energy must increase. But in the nonequilibrium thermodynamics of open structures, we are not

252

concerned with the absolute value of a system's total free energy as much as with the free energy density. (After all, a galaxy clearly has more energy than a cell, but of course galaxies also have greater sizes and masses; it is the free energy *density* that characterizes organization just as it was the radiation energy density and matter energy density that were important earlier in the Universe.) In fact, what is most important is the *rate* at which free energy enters a system of some given size. In other words, the operative quantity used to specify the order and organization in any system is the flux of free energy density, denoted here by the symbol $\mathscr{F}$.

I have accordingly computed the energy fluxing through various structures representing a wide spectrum of order or information content; knowing the size and scale of such structures, I then estimated the free energy flux density, $\mathscr{F}$, which should be a measure of the energy available to order a given amount of matter. For example, consider a star, in particular an average star such as our Sun. The solar luminosity is 4×10^{33} ergs s^{-1}, while its mass is 2×10^{33} grams; therefore, $\mathscr{F}$ for the Sun equals 2 erg s^{-1} gm^{-1}. A typical galaxy would have about half a star's $\mathscr{F}$ since roughly half the mass of a galaxy is in the form of stars. By contrast, planets are more complex, and not surprisingly their $\mathscr{F}$'s are larger. Take the amount of energy needed to order the Earth's climasphere: The total solar flux reaching Earth is 1.8×10^{24} erg s^{-1}, of which only 70 percent penetrates the atmosphere (since Earth's albedo is 0.3); as our planet's air totals 5×10^{18} kgm and its oceans weigh roughly double that, the value of $\mathscr{F}$ for planet Earth is 80 ergs s^{-1} gm^{-1}.

Living systems require substantially larger values of $\mathscr{F}$ to maintain their order. Plants, for instance, require 17 kJ for each gram of photosynthesizing biomass. Since the annual production rate of Earth's biosphere is 1.7×10^{17} gm yr^{-1}, the entire biosphere must use energy at the rate of about 10^{21} ergs s^{-1}. And given that the total mass of the terrestrial biomass has been estimated at 2×10^{18} gm, then $\mathscr{F}$ for the physico-chemical process of photosynthesis is some 500 ergs s^{-1} gm^{-1}. Humans, by contrast, needing roughly 2,500 kcal day^{-1} (or 120 watts in the form of food) and having an average body mass of 70 kgm (or some 150 pounds), have an $\mathscr{F}$ of 1.7×10^4 ergs s^{-1} gm^{-1}. And the most exquisitely complex clump of matter in the known Universe, the human brain with a typical mass of 1,300 grams, requires about 40 kcal day^{-1} (or 20 watts) to function properly; our brains therefore have an $\mathscr{F}$ value of 1.5×10^5 ergs s^{-1} gm^{-1}.

The table below summarizes these representative values of $\mathscr{F}$, expressed in units of ergs s^{-1} gm^{-1}, for again I stress that $\mathscr{F}$ is an energy flux (i.e., per unit time) "density" (i.e., per unit mass).

A TABLE OF FREE ENERGY FLUX DENSITIES

Structure	$\mathscr{F}$ (ergs s^{-1} gm^{-1})
Milky Way Galaxy	1
Sun	2
Earth's climasphere	80
Earth's biosphere (plants)	500
human body (animals)	17,000
human brain	150,000

Clearly $\mathscr{F}$ increases dramatically as more intricately ordered structures have emerged throughout history. Though the flux of energy through a star is obviously hugely larger than that through our human bodies or brains, the flux *densities* are much larger for the latter. Thus, organized systems observed in the Universe do in fact increase in information content, in actuality $\mathscr{F}$ faster than I doubtlessly because factors such as biological evolution can direct additional free energy over and above that provided by the expansion of the Universe.

Finally, I have graphed below a superposition of the temporal dependence of aT^4, $\rho_m c^2$, and $\mathscr{F}$, thus signifying the variation of energy, matter, and organization throughout all history. (Note that this comparison, at this point in an ongoing research program, is only aesthetic since $\mathscr{F}$ has units different from those of the other two energy densities.) Whereas the previously discussed eclipse of radiation energy density by matter energy density denoted the onset of the Matter Era, the ascent of clustered information over both these energy densities heralds the Life Era. In particular, as noted earlier, toward the end of Chapter 3, the controlled use of radiation by animated structures (i.e., when the free energy density exceeds the radiation energy density, $\mathscr{F} > aT^4$, beginning with a simple cell), we identify with one of the first great inventions of biological evolution: photosynthesis. This is a grand event wherein life dominates radiation, utilizing sunlight in a survival-related fashion. An even grander event occurs when life forms dominate matter (i.e., when the free energy density exceeds the matter energy density, $\mathscr{F} > \rho_m c^2$, beginning with technological intelligence on Earth and possibly elsewhere). *Only the latter of these two changes heralds a genuinely new era*—herein called the Life Era—because only with the origin of technologically manipulative life, not just life itself, does life exert leverage over both radiation *and* matter.

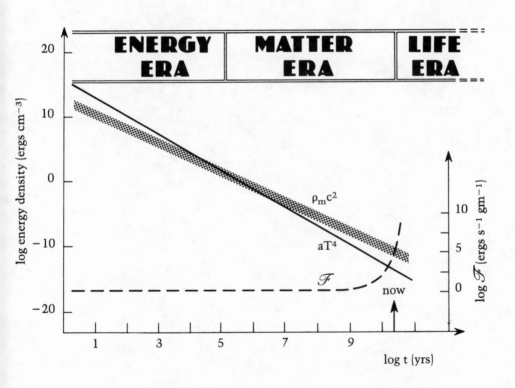

For Further Reading

The following general sources are representative of the many works I found useful while writing the present book; they thus comprise suitable material for further reading:

General Cosmic Evolution

E. Chaisson, *Cosmic Dawn*, Atlantic-Little, Brown, 1981.

H. Reeves, *Atoms of Silence*, MIT Press, 1984.

C. Sagan, *Cosmos*, Random House, 1980.

G. A. Seielstad, *Cosmic Ecology*, University of California Press, 1983.

H. Shapley, *Of Stars and Men*, Beacon, 1958.

History of Change

H. J. Birx, *Theories of Evolution*, Thomas, 1984.

D. J. Boorstin, *The Discoverers*, Random House, 1983.

W. Durant, *The Story of Philosophy*, Simon & Schuster, 1933.

E. Mayr, *The Growth of Biological Thought*, Harvard University Press, 1982.

B. Russell, *A History of Western Philosophy*, Simon & Schuster, 1945.

Physics of Change

H. F. Blum, *Time's Arrow and Evolution*, Princeton University Press, 1968.

J. Campbell, *Grammatical Man*, Simon & Schuster, 1982.

H. Haken, *Science of Structure: Synergetics*, Van Nostrand, 1981.

J. Monod, *Chance and Necessity*, Knopf, 1971.

I. Prigogine and I. Stengers, *Order Out of Chaos*, Bantam, 1984.

Two Preeminent Changes

M. Eigen and R. Winkler, *The Laws of the Game*, Knopf, 1981.

B. Gal-Or, *Cosmology, Physics, and Philosophy*, Springer-Verlag, 1981.

E. Jantsch, ed., *The Evolutionary Vision*, Westview, 1981.

H. R. Pagels, *Perfect Symmetry*, Simon & Schuster, 1985.

J. S. Trefil, *The Moment of Creation*, Scribner's, 1983.

Implications for the Life Era

J. Billingham, ed., *Life in the Universe*, MIT Press, 1981.

J. S. Huxley, *Evolutionary Ethics*, Oxford University Press, 1943.

E. Laszlo, *Evolution*, New Science Library, 1987.

J. Salk, *Anatomy of Reality*, Columbia University Press, 1983.

P. Teilhard de Chardin, *The Phenomenon of Man*, Harper, 1959.

Cosmic Life and Consciousness

J. R. Clark, *The Great Living System*, Skinner House, 1977.

J. E. Lovelock, *Gaia*, Oxford University Press, 1979.

E. Schrödinger, *What Is Life?* and *Mind & Matter*, Cambridge University Press, 1967.

L. Thomas, *The Lives of a Cell*, Viking, 1974.

G. Wald, "Life and Mind in the Universe," *International Journal of Quantum Chemistry* 11 (1984):1.

Appendix

F. J. Dyson, "Time Without End," *Reviews of Modern Physics* 51, (1979):447.

H. J. Morowitz, *Energy Flow in Biology*, Academic, 1968.

P. Morrison, "Thermodynamics and the Origin of Life," in *Molecules in the Galactic Environment*, M. Gordon and L. Snyder, eds., Wiley, 1973.

A. R. Peacocke, *The Physical Chemistry of Biological Organization*, Oxford University Press, 1983.

M. Taub, *Evolution of Matter and Energy*, Springer, 1985.

About
the Author

Eric J. Chaisson has published more than a hundred scientific articles, most of them in the professional journals. He has also authored several books, including *Cosmic Dawn*, which won the Phi Beta Kappa Prize, the American Institute of Physics Award, and an American Book Award Nomination for distinguished science writing. Trained initially in condensed-matter physics, Chaisson received his doctorate in astrophysics from Harvard University, where he spent a decade as a member of the faculty of Arts and Sciences. While at the Harvard-Smithsonian Center for Astrophysics, he won fellowships from the National Academy of Sciences and the Sloan Foundation, as well as Harvard's Bok and Smith prizes. Dr. Chaisson is currently a research physicist at MIT's Lincoln Laboratory and teaches astrophysics at Harvard and Wellesley colleges. He lives with his wife and two children on the common in the village of Harvard, Massachusetts.